AF380480

Methods in Molecular Biology

Series Editor
John M. Walker
School of Life and Medical Sciences
University of Hertfordshire
Hatfield, Hertfordshire, AL10 9AB, UK

For further volumes:
http://www.springer.com/series/7651

Brain Tumor Stem Cells

Methods and Protocols

Edited by

Sheila K. Singh

Stem Cell and Cancer Research Institute, McMaster University, Hamilton, ON, Canada

Chitra Venugopal

Stem Cell and Cancer Research Institute, McMaster University, Hamilton, ON, Canada

 Humana Press

Editors
Sheila K. Singh
Stem Cell and Cancer
Research Institute
McMaster University
Hamilton, ON, Canada

Chitra Venugopal
Stem Cell and Cancer
Research Institute
McMaster University
Hamilton, ON, Canada

ISSN 1064-3745　　　　ISSN 1940-6029　(electronic)
Methods in Molecular Biology
ISBN 978-1-4939-8804-4　　　ISBN 978-1-4939-8805-1　(eBook)
https://doi.org/10.1007/978-1-4939-8805-1

Library of Congress Control Number: 2018957127

This Humana Press imprint is published by the registered company Springer Science+Business Media, LLC, part of Springer Nature.
The registered company address is: 233 Spring Street, New York, NY 10013, U.S.A.

Preface

The study of human cancer stem cell (CSC) populations is fraught with technical challenges and seemingly insurmountable experimental tribulations, including limited and highly precious patient-derived samples, low cell numbers, optimal growth conditions, and difficulties in sorting and purifying heterogeneous cell populations derived from cancer stem cell hierarchies. And yet the rewards of implementing the newest and most advanced technologies to obtain novel data from these CSCs far exceed any of the challenges faced by scientists, as discoveries made in human CSC populations are one step closer to translation into new therapeutic options for cancer patients.

In this textbook, we have carefully curated and compiled the best methodologies and experimental techniques to profile and extract maximal data from brain tumor stem cells (BTSCs), the experimental paradigm for brain cancer research that offers insights into CSC populations that may drive not only tumor initiation but also tumor recurrence and patient relapse. The BTSC model recapitulates scientific observations made in brain cancer patients, and we seek to give the reader a comprehensive understanding of the skills and techniques that will unlock data from this most informative subset of cells.

We do hope that you will enjoy reading these succinct methodological briefs on the most innovative technologies and applications to BTSC research, provided by international experts in the field. Armed with the best available knowledge of BTSC techniques, we wish you happy data acquisition on your journey to better understand the complexities of brain cancer.

Hamilton, ON, Canada

Sheila K. Singh
Chitra Venugopal

Contents

Preface .... *v*

Contributors .... *ix*

1 Introduction to Brain Tumor Stem Cells 1
   Nicolas Yelle, David Bakhshinyan, Chitra Venugopal,
   and Sheila K. Singh

2 Isolation and Culture of Glioblastoma Brain Tumor Stem Cells 11
   Charles Chesnelong, Ian Restall, and Samuel Weiss

3 Establishment and Culture of Patient-Derived Primary
   Medulloblastoma Cell Lines 23
   Sara Badodi, Silvia Marino, and Loredana Guglielmi

4 Bioinformatic Strategies for the Genomic and Epigenomic
   Characterization of Brain Tumors 37
   Vijay Ramaswamy and Michael D. Taylor

5 Detecting Stem Cell Marker Expression Using the NanoString
   nCounter System 57
   Scott Ryall, Anthony Arnoldo, Javal Sheth, Sheila K. Singh,
   and Cynthia Hawkins

6 Flow Cytometric Analysis of Brain Tumor Stem Cells 69
   Minomi K. Subapanditha, Ashley A. Adile, Chitra Venugopal,
   and Sheila K. Singh

7 In Vitro Self-Renewal Assays for Brain Tumor Stem Cells 79
   Mathieu Seyfrid, David Bobrowski, David Bakhshinyan, Nazanin Tatari,
   Chitra Venugopal, and Sheila K. Singh

8 Differentiation of Brain Tumor Initiating Cells 85
   Michelle M. Kameda-Smith, Minomi K. Subapanditha, Sabra K. Salim,
   Chitra Venugopal, and Sheila K. Singh

9 The Study of Brain Tumor Stem Cell Migration 93
   Montserrat Lara-Velazquez, Rawan Al-kharboosh, Luis Prieto,
   Paula Schiapparelli, and Alfredo Quiñones-Hinojosa

10 The Study of Brain Tumor Stem Cell Invasion 105
    Rawan Al-kharboosh, Montserrat Lara-Velazquez,
    Luis Prieto, Rachel Sarabia-Estrada, and Alfredo Quiñones-Hinojosa

11 Cell Cycle Dynamics in Glioma Cancer Stem Cells 117
    Ingrid Qemo and Lisa A. Porter

12 Embryonic Stem Cell Models of Human Brain Tumors 127
    Ludivine Coudière Morrison, Nazanin Tatari,
    and Tamra E. Werbowetski-Ogilvie

viii   Contents

13 Chromatin Immunoprecipitation (ChIP) Protocols for the Cancer
    and Developmental Biology Laboratory 143
    Hunter McColl, Jamie L. Zagozewski, and David D. Eisenstat

14 EPH Profiling of BTIC Populations in Glioblastoma Multiforme
    Using CyTOF 155
    *Amy X. Hu, Jarrett J. Adams, Parvez Vora, Maleeha Qazi, Sheila K. Singh,
    Jason Moffat, and Sachdev S. Sidhu*

15 Pooled Lentiviral CRISPR-Cas9 Screens for Functional Genomics
    in Mammalian Cells 169
    *Michael Aregger, Megha Chandrashekhar, Amy Hin Yan Tong,
    Katherine Chan, and Jason Moffat*

16 In Vitro Assays for Screening Small Molecules 189
    *Ashley A. Adile, David Bakhshinyan, Chitra Venugopal,
    and Sheila K. Singh*

17 Drug Delivery in an Orthotopic Tumor Stem Cell-Based Model
    of Human Glioblastoma 197
    Elena Binda, Alberto Visioli, Nadia Trivieri, and Angelo Luigi Vescovi

18 Engineering Inducible Knock-In Mice to Model Oncogenic Brain
    Tumor Mutations from Endogenous Loci 207
    Jon D. Larson and Suzanne J. Baker

19 In Vivo Murine Models of Brain Metastasis 231
    Mohini Singh, Neil Savage, and Sheila K. Singh

20 Cellular Magnetic Resonance Imaging for Tracking Metastatic
    Cancer Cells in the Brain 239
    *Katie M. Parkins, Ashley V. Makela, Amanda M. Hamilton,
    and Paula J. Foster*

Index *253*

Contributors

JARRETT J. ADAMS • *Department of Molecular Genetics, University of Toronto, Toronto, ON, Canada; Donnelly Centre for Cellular and Biomolecular Research, University of Toronto, Toronto, ON, Canada*

ASHLEY A. ADILE • *McMaster Stem Cell and Cancer Research Institute, McMaster University, Hamilton, ON, Canada; Department of Biochemistry and Biomedical Sciences, Faculty of Health Sciences, McMaster University, Hamilton, ON, Canada*

RAWAN AL-KHARBOOSH • *Department of Neurosurgery, Mayo Clinic Florida, Jacksonville, FL, USA; Mayo Clinic Graduate School, Mayo Clinic College of Medicine, Rochester, MN, USA*

MICHAEL AREGGER • *Donnelly Centre for Cellular and Biomolecular Research, University of Toronto, Toronto, ON, Canada*

ANTHONY ARNOLDO • *Department of Pediatric Laboratory Medicine, Hospital for Sick Children, Toronto, ON, Canada*

SARA BADODI • *Blizard Institute, Barts and The London School of Medicine and Dentistry, Queen Mary University of London, London, UK*

SUZANNE J. BAKER • *Department of Developmental Neurobiology, St. Jude Children's Research Hospital, Memphis, TN, USA*

DAVID BAKHSHINYAN • *McMaster Stem Cell and Cancer Research Institute, McMaster University, Hamilton, ON, Canada; Department of Biochemistry and Biomedical Sciences, McMaster University, Hamilton, ON, Canada*

ELENA BINDA • *Cancer Stem Cells Unit, Institute for Stem Cell Biology, Regenerative Medicine and Innovative Therapies–ISBReMIT, IRCSS, S. Giovanni Rotondo, FG, Italy*

DAVID BOBROWSKI • *McMaster Stem Cell and Cancer Research Institute, McMaster University, Hamilton, ON, Canada; Faculty of Health Sciences, McMaster University, Hamilton, ON, Canada*

KATHERINE CHAN • *Donnelly Centre for Cellular and Biomolecular Research, University of Toronto, Toronto, ON, Canada*

MEGHA CHANDRASHEKHAR • *Donnelly Centre for Cellular and Biomolecular Research, University of Toronto, Toronto, ON, Canada; Department of Molecular Genetics, University of Toronto, Toronto, ON, Canada*

CHARLES CHESNELONG • *Department of Cell Biology and Anatomy, Faculty of Medicine, University of Calgary, Calgary, AB, Canada; Department of Physiology and Pharmacology, Faculty of Medicine, University of Calgary, Calgary, AB, Canada; Hotchkiss Brain Institute, University of Calgary, Calgary, AB, Canada*

DAVID D. EISENSTAT • *Department of Medical Genetics, University of Alberta, Edmonton, AB, Canada; Department of Pediatrics, University of Alberta, Edmonton, AB, Canada; Department of Oncology, University of Alberta, Edmonton, AB, Canada*

PAULA J. FOSTER • *Robarts Research Institute, Western University, London, ON, Canada*

LOREDANA GUGLIELMI • *Blizard Institute, Barts and The London School of Medicine and Dentistry, Queen Mary University of London, London, UK*

AMANDA M. HAMILTON • *Robarts Research Institute, Western University, London, ON, Canada*

CYNTHIA HAWKINS • *Division of Cell Biology, Hospital for Sick Children Research Center, Toronto, ON, Canada; Department of Laboratory Medicine and Pathobiology, Faculty of Medicine, University of Toronto, Toronto, ON, Canada; Department of Pediatric Laboratory Medicine, Hospital for Sick Children, Toronto, ON, Canada*

AMY X. HU • *Department of Molecular Genetics, University of Toronto, Toronto, ON, Canada; Donnelly Centre for Cellular and Biomolecular Research, University of Toronto, Toronto, ON, Canada*

MICHELLE M. KAMEDA-SMITH • *McMaster Stem Cell and Cancer Research Institute, McMaster University, Hamilton, ON, Canada*

MONTSERRAT LARA-VELAZQUEZ • *Department of Neurosurgery, Mayo Clinic Florida, Jacksonville, FL, USA; Plan of Combined Studies in Medicine, Faculty of Medicine, National Autonomous University of Mexico, Mexico City, Mexico*

JON D. LARSON • *Department of Developmental Neurobiology, St. Jude Children's Research Hospital, Memphis, TN, USA*

ASHLEY V. MAKELA • *Robarts Research Institute, Western University, London, ON, Canada*

SILVIA MARINO • *Blizard Institute, Barts and The London School of Medicine and Dentistry, Queen Mary University of London, London, UK*

HUNTER MCCOLL • *Department of Medical Genetics, University of Alberta, Edmonton, AB, Canada*

JASON MOFFAT • *Donnelly Centre for Cellular and Biomolecular Research, University of Toronto, Toronto, ON, Canada; Department of Molecular Genetics, University of Toronto, Toronto, ON, Canada; Canadian Institute for Advanced Research, Toronto, ON, Canada*

LUDIVINE COUDIÈRE MORRISON • *Regenerative Medicine Program, Department of Biochemistry and Medical Genetics, University of Manitoba, Winnipeg, MB, Canada; Regenerative Medicine Program, Department of Physiology and Pathophysiology, University of Manitoba, Winnipeg, MB, Canada*

KATIE M. PARKINS • *Robarts Research Institute, Western University, London, ON, Canada*

LISA A. PORTER • *Department of Biological Sciences, University of Windsor, Windsor, ON, Canada*

LUIS PRIETO • *Mayo Clinic Graduate School, Mayo Clinic College of Medicine, Rochester, MN, USA*

MALEEHA QAZI • *McMaster Stem Cell and Cancer Research Institute, McMaster University, Hamilton, ON, Canada; Department of Biochemistry and Biomedical Sciences, McMaster University, Hamilton, ON, Canada*

INGRID QEMO • *Department of Biological Sciences, University of Windsor, Windsor, ON, Canada*

ALFREDO QUIÑONES-HINOJOSA • *Department of Neurosurgery, Mayo Clinic Florida, Jacksonville, FL, USA*

VIJAY RAMASWAMY • *Division of Haematology/Oncology, Hospital for Sick Children, Toronto, ON, Canada; Program in Developmental and Stem Cell Biology, Hospital for Sick Children, Toronto, ON, Canada*

IAN RESTALL • *Department of Cell Biology and Anatomy, Faculty of Medicine, University of Calgary, Calgary, AB, Canada; Hotchkiss Brain Institute, University of Calgary, Calgary, AB, Canada*

SCOTT RYALL • *Division of Cell Biology, Hospital for Sick Children Research Center, Toronto, ON, Canada; Department of Laboratory Medicine and Pathobiology, Faculty of Medicine, University of Toronto, Toronto, ON, Canada*

SABRA K. SALIM • *McMaster Stem Cell and Cancer Research Institute, McMaster University, Hamilton, ON, Canada*

RACHEL SARABIA-ESTRADA • *Department of Neurosurgery, Mayo Clinic Florida, Jacksonville, FL, USA*

NEIL SAVAGE • *Stem Cell and Cancer Research Institute, McMaster University, Hamilton, ON, Canada; Department of Surgery, McMaster University, Hamilton, ON, Canada*

PAULA SCHIAPPARELLI • *Department of Neurosurgery, Mayo Clinic Florida, Jacksonville, FL, USA*

MATHIEU SEYFRID • *McMaster Stem Cell and Cancer Research Institute, McMaster University, Hamilton, ON, Canada*

JAVAL SHETH • *Division of Cell Biology, Hospital for Sick Children Research Center, Toronto, ON, Canada; Department of Pediatric Laboratory Medicine, Hospital for Sick Children, Toronto, ON, Canada*

SACHDEV S. SIDHU • *Department of Molecular Genetics, University of Toronto, Toronto, ON, Canada; Donnelly Centre for Cellular and Biomolecular Research, University of Toronto, Toronto, ON, Canada*

MOHINI SINGH • *Stem Cell and Cancer Research Institute, McMaster University, Hamilton, ON, Canada; Department of Biochemistry and Biomedical Sciences, McMaster University, Hamilton, ON, Canada*

SHEILA K. SINGH • *Stem Cell and Cancer Research Institute, McMaster University, Hamilton, ON, Canada; Department of Biochemistry and Biomedical Sciences, McMaster University, Hamilton, ON, Canada; Department of Surgery, McMaster University, Hamilton, ON, Canada*

MINOMI K. SUBAPANDITHA • *McMaster Stem Cell and Cancer Research Institute, McMaster University, Hamilton, ON, Canada; Department of Biochemistry and Biomedical Sciences, Faculty of Health Sciences, McMaster University, Hamilton, ON, Canada*

NAZANIN TATARI • *Regenerative Medicine Program, Department of Biochemistry and Medical Genetics, University of Manitoba, Winnipeg, MB, Canada; Regenerative Medicine Program, Department of Physiology and Pathophysiology, University of Manitoba, Winnipeg, MB, Canada; McMaster Stem Cell and Cancer Research Institute, McMaster University, Hamilton, ON, Canada; Department of Biochemistry and Biomedical Sciences, McMaster University, Hamilton, ON, Canada*

MICHAEL D. TAYLOR • *Division of Neurosurgery, Hospital for Sick Children, Toronto, ON, Canada; Program in Developmental and Stem Cell Biology, Hospital for Sick Children, Toronto, ON, Canada*

AMY HIN YAN TONG • *Donnelly Centre for Cellular and Biomolecular Research, University of Toronto, Toronto, ON, Canada*

NADIA TRIVIERI • *Cancer Stem Cells Unit, Institute for Stem Cell Biology, Regenerative Medicine and Innovative Therapies–ISBReMIT, IRCSS, S. Giovanni Rotondo, FG, Italy*

CHITRA VENUGOPAL • *Stem Cell and Cancer Research Institute, McMaster University, Hamilton, ON, Canada; Department of Surgery, Faculty of Health Sciences, McMaster University, Hamilton, ON, Canada*

ANGELO LUIGI VESCOVI • *Cancer Stem Cells Unit, Institute for Stem Cell Biology, Regenerative Medicine and Innovative Therapies–ISBReMIT, IRCSS, S. Giovanni Rotondo, FG, Italy; Department of Biotechnology and Biosciences, University of Milan Bicocca, Milan, Italy; Hyperstem SA, Lugano, Switzerland*

ALBERTO VISIOLI • *StemGen SpA, Milan, Italy*

PARVEZ VORA • *McMaster Stem Cell and Cancer Research Institute, McMaster University, Hamilton, ON, Canada; Department of Biochemistry and Biomedical Sciences, McMaster University, Hamilton, ON, Canada*

SAMUEL WEISS • *Department of Cell Biology and Anatomy, Faculty of Medicine, University of Calgary, Calgary, AB, Canada; Hotchkiss Brain Institute, University of Calgary, Calgary, AB, Canada*

TAMRA E. WERBOWETSKI-OGILVIE • *Regenerative Medicine Program, Department of Biochemistry and Medical Genetics, University of Manitoba, Winnipeg, MB, Canada; Regenerative Medicine Program, Department of Physiology and Pathophysiology, University of Manitoba, Winnipeg, MB, Canada*

NICOLAS YELLE • *Stem Cell and Cancer Research Institute, McMaster University, Hamilton, ON, Canada; Department of Biochemistry and Biomedical Sciences, McMaster University, Hamilton, ON, Canada*

JAMIE L. ZAGOZEWSKI • *Department of Medical Genetics, University of Alberta, Edmonton, AB, Canada; Department of Biochemistry and Medical Genetics, University of Manitoba, Winnipeg, MB, Canada*

Chapter 1

Introduction to Brain Tumor Stem Cells

Nicolas Yelle, David Bakhshinyan, Chitra Venugopal, and Sheila K. Singh

Abstract

From stem cells, to the cancer stem cell hypothesis and intratumoral heterogeneity, the following introductory chapter on brain tumor stem cells explores the history of normal and cancerous stem cells, and their implication in the current model of brain tumor development. The origins of stem cells date back to the 1960s, when they were first described as cells capable of self-renewal, extensive proliferation, and differentiation. Since then, many advances have been made and adult stem cells are now known to be present in a very wide variety of tissues. Neural stem cells were subsequently discovered 30 years later, which was shortly followed by the discovery of cancer stem cells in leukemia and in brain tumors over the next decade, effectively enabling a new understanding of cancer. Since then, many markers including CD133, brain cancer stem cells have been implicated in a variety of phenomena including intratumoral heterogeneity on the genomic, cellular, and functional levels, tumor initiation, chemotherapy-resistance, radiation-resistance, and are believed to be ultimately responsible for tumor relapse. Understanding this small and rare population of cells could be the key to solving the great enigma that is cancer.

Key words Neural stem cells, Cancer stem cells, Brain tumor stem cells, Intratumoral heterogeneity, Brain tumor initiating cell makers

---

1 Stem Cells

The human body contains such an unbelievable amount of phenotypically and functionally distinct cells that it is sometimes hard to believe they all carry the same genetic information and arise from a single totipotent cell. Through a series of cellular differentiation, proliferation, and self-renewal, the zygote will generate every known human cell, including the various stem cells, which have preserved the stated abilities.

Stem cells were originally discovered in the 1960s by James Till and Ernest McCulloch after they performed modified spleen colony-formation assays in vivo [1–3]. Stem cells were described and characterized by quantifying the formation of spleen colonies after mouse donor bone marrow cells were transplanted into irradiated mice. Additionally, they established that stem cells are to be studied in vitro based on their functional properties of self-renewal,

Sheila K. Singh and Chitra Venugopal (eds.), *Brain Tumor Stem Cells: Methods and Protocols*, Methods in Molecular Biology, vol. 1869, https://doi.org/10.1007/978-1-4939-8805-1_1, © Springer Science+Business Media, LLC, part of Springer Nature 2019

extensive proliferation, and multi-lineage differentiation, and in vivo based on their colony-forming abilities. More recently, the definition of stem cells has been amended to self-renewing and highly proliferative undifferentiated cells that are flexible in their abilities to produce differentiated progenitor cells and regenerate injured tissue [4]. Stem cells were initially thought to be limited to tissues with high cellular turnover such as blood, skin or intestines, but have since been shown to be present in most tissues including breast, brain, and lungs.

There are many different types of stem cells, each with differences in their stem abilities, and each giving rise to different cells. Embryonic stem cells (ESCs) are found in the blastocyst, and are pluripotent—they can give rise to all tissues, except for the placenta and umbilical cord. Adult or somatic stem cells are multipotent tissue-specific stem cells. These cells are more specialized than ESCs as they can only give rise to a limited number of lineages. Adult stem cells are found in small populations throughout the juvenile and adult body, and consist of many different types including hematopoietic stem cells (HSCs), mesenchymal stem cells (MSCs), and neural stem cells (NSCs), among many more.

2   Neural Stem Cells

Neural stem cells (NSCs) are the stem cells of the nervous system, with the ability to give rise to the neural and glial cells of the nervous system. The pioneering studies of NSCs have led to their successful isolation in both the embryonic central nervous system (CNS) [5–8] and the peripheral nervous system (PNS) [9]. It was not until 1992, when a stem-like cell capable of self-renewal and of generating progenitor cells with a restricted lineage potential were isolated from the adult brain [10, 11]. The major limitation of the studies aiming to identify markers of NSCs lies in the fact that NSCs can only be defined retrospectively based on their functional properties [12–14]. When cultured adherently, NSCs can give rise to colonies containing undifferentiated cells through self-renewal, as well as differentiated cells such as neurons, oligodendrocytes, and glial cells through the process of differentiation. However when grown in suspension cultures, NSCs form neurospheres, allowing researchers to utilize assays quantifying the number of neurospheres formed from a single cell suspension to estimate the frequency of self-renewing units in culture after various perturbations.

The methodology utilized to isolate NCSs has been modified and improved since the early studies, and a variety of protocols now exist. Despite their difference, all protocols follow the same three major steps: (1) tissue dissection, (2) digestion of the extracellular matrix, and (3) cell culture-based enrichment and expansion of NSC population. Because NSCs and neural progenitor cells

(NPCs) with varying proliferative capacities exist in different parts of CNS, it is important to consider whether macrodissection, isolating a specific area of interest, is required prior to a finer microdissection and mincing the tissue into small pieces [15]. To isolate NSCs from the surrounding extracellular matrix, a protease is required. Although protocols utilizing various proteases including trypsin [7], papain [16], collagenase [17], dispase [16], and hyaluronidase [18] have shown efficacy in isolating NSCs, they often vary in the incubation time and are frequently combined for a more complete isolation. The use of proteases often results in cell lysis and a subsequent increase in concentration of free DNA, making the use of Deoxyribonuclease I (DNAse I) a common feature of NSC isolation protocols. To ensure a single cell solution post-treatment with proteases, several mechanical disaggregation techniques can be utilized, such as the use of pipette tips, sterile syringes, and cell microfilters with varying pore sizes. Once a single cell suspension has been obtained, different cell culturing methods can be used to isolate a specific subpopulation. Alternatively, a cell surface marker-based separation can be desired and achieved through fluorescence activated cell sorting (FACS).

3 Cancer Stem Cells

The cancer stem cell (CSC) hypothesis is a theorem stating that a very small population of cells with stem cell-like properties of self-renewal, extensive proliferation, and differentiation drive tumor initiation, formation, and maintenance. This hypothesis has been around for some time, and saw its true beginnings in the latter part of the twentieth century as a result of germ cell cancer, or teratocarcinoma research. Researchers first observed not only that teratocarcinomas had similar cellular components to normal healthy tissues (differentiated, progenitor, and stem cells), but also that malignant cells were exclusively found in embryoid bodies (i.e., ESC aggregates) [19]. It was later found that only these were able to propagate in vitro and form tumor after transplantation in vivo [20, 21]. From this point on researchers began focusing their attention on identifying these elusive cancer stem cells.

CSCs were first discovered in 1997 by Dr. John Dick in acute myeloid leukemia. Leukemic stem cells (LSCs), or SCID leukemia initiating cells (SL-ICs), were identified thanks to previous studies which found that the CD34+/CD38- population of HSCs had cytogenetic abnormalities [22, 23]. LSCs were then identified and characterized through serial transplantation into immunocompromised mice, based on their proliferative, self-renewing, and differentiation abilities. The rare CD34+/CD38- population was shown to be able to form the entire leukemic hierarchy, just like an HSC can form the healthy hematopoietic hierarchy [24]. Bonnet and

Dick were also able to demonstrate that LSCs and healthy HSCs shared many similarities including cell surface markers, and their mostly quiescent nature, providing further evidence that HSCs are at the origin of the disease, hence further strengthening the CSC hypothesis [25, 26].

The first solid tumor CSCs were subsequently discovered in 2003 by Dr. Michael Clarke in breast tumors [27]. Similarly to the discovery of LSCs, breast cancer stem cells (BCSCs) were identified using the markers CD44+/CD24-, which mark a population with elevated stem-like, adhesion, migration, and invasion properties [27–30]. The discovery of BCSCs was shortly followed by the discovery of brain tumor stem cells, also known as brain tumor initiating cells (BTICs) [31, 32], and others including pancreatic [33], colon [34, 35], and colorectal [36] cancer stem cells. Nowadays, CSCs are known to exist in many solid tumors including breast, brain, head and neck, lung, skin, pancreatic, liver, prostate, colon cancer, and more.

Over the past 15 years, our knowledge of the tumor landscape has notably changed as we now know solid tumors are not clonal populations, but are in fact subject to significant intratumoral heterogeneity (ITH) on the functional, cellular, and genetic levels. This phenomenon can be explained by the CSC hypothesis, which states that tumors are initiated by malignant cells with stem-like properties which allow them to differentiate and proliferate extensively in a hierarchal manner, leading to the vast observed heterogeneity. However, whether tumor-initiating cells originate from malignant stem cells, or from the acquisition of stem-like properties via malignant transformations and dedifferentiation, remains unknown. In solid tumors, ITH and CSCs have been linked to treatment resistance and tumor relapse. The heterogeneous landscape of a solid tumor is able to bring about functionally different cells, some of which will be able to resist chemotherapeutic drugs and radiation to varying degrees [37–42]. Typically, conventional therapies will target the bulk tumor mass of more highly proliferative cells, which leaves quiescent cells, such as the majority of healthy and cancer stem cells, unaffected. In fact, CSCs have been shown to have heightened chemotherapy and radiation therapy resistance properties [43–45], further corroborating the CSC hypothesis and its role in tumor progression and recurrence. Furthermore, CSCs have been shown to resort to symmetric division at a higher rate when submitted to cellular stress such as chemotherapy and radiation, thus increasing the tumor initiating population of tumor cells in the recurrent tumor [46]. This can explain why recurrences are often more aggressive than the initial tumor, and why they are resistant to conventional therapy. Due to these facts, many have made the case against monotherapies, explaining why such approaches are doomed to fail [47–49].

Despite the many advances accomplished in the field, the CSC hypothesis remains controversial, as not all cancers seem to be hierarchal, there is a lack of universal CSC markers, and the CSCs themselves are hard to properly define, leading to some doubting their very existence [50, 51]. Currently, CSCs are defined as capable of initiating and propagating tumors, and leading to hierarchal tumors, and being capable of extensive self-renewal and proliferation [52, 53]. Although lacking universal markers, CSCs are also identified by the presence of specific stem cell markers based on the tissue of origin, and the absence of lineage-specific markers based on the standards established by the Dick and Clarke groups [27, 54].

4 Brain Tumor Stem Cells

The discovery of breast CSCs was shortly followed by the discovery and isolation of brain or neural CSCs, also known as brain tumor initiating cells (BTICs) in the major brain tumor types including gliomas, medulloblastomas, and ependymomas [31, 32, 55–58]. BTICs were discovered after various groups cultured human adult and pediatric brain tumors in serum-free NSC-enrichment media [31, 55–57]. Further experiments using fluorescence and magnetic cell sorting techniques were used to isolate a malignant population possessing enhanced proliferation, self-renewal, and differentiation properties in vitro and in vivo, based on the well-established NSC marker CD133 (Prominin-1) [31, 32]. Singh and colleagues additionally demonstrated that the CD133+ populations were not only able to give rise to self-renewing colonies (termed neurospheres) in vitro, but that they also form tumors in mouse xenografts at much higher rates than the CD133- cell populations. Additionally, tumors derived from CD133+ cells were heterogeneous and a virtual phenocopy of the patient's original tumor, presenting a good animal model for the disease and suggesting the presence of a cellular hierarchy originating from CD133+ cell fraction [32]. Being CSCs, these cells have also been shown to be resistant to chemotherapy [43], and radiation therapy [44, 45]. The collective roles brain CSCs have been shown to play in tumor initiation, maintenance, and in their ability to escape therapy and cause recurrence makes them a significant biological target for therapeutic development, making in vitro and in vivo brain tumor stem cell models pertinent platforms for future drug discovery. Reinforcing the potential and importance of these models is the fact that tumors bearing a heightened CD133 expression have been shown to be indicative of poorer patient prognosis [59–62]. Since the discovery of CD133+ BTICs, other markers including known NSC markers have been shown to identify other BTIC populations including CD15 [63], ITGA6 [64], L1CAM

[65], EphA2 [66], EphA3 [67], EphB2 [68], Sox2 [69], Oct4 [70], and Msi1 [71]. Other genes important in preserving a stem cell-like state such as Bmi1 [56, 72] and FoxG1 [73] have also been shown to have an increased expression in BTICs.

Brain CSCs can be isolated and purified through fluorescence or magnetic activated cell sorting using BTIC surface markers. The antibodies directed against the BTIC surface markers are either labeled with fluorophores or magnetic nanoparticles. Although both methods will effectively sort tumor populations into BTIC and non-BTIC fractions, there are unique advantages and disadvantages associated with each method. Fluorescence activated cell sorting (FACS) allows for higher purity of the fractionated cell population as each antibody-labeled cell is individually analyzed for fluorescence and sorted according to predetermined parameters including cell size, granularity, and viability. However, shear forces due to pressurized fluids may reduce the post-sort viability of cells [74]. Cells sorted using magnetic activated cell sorting (MACS) are not individually analyzed for marker expression, resulting in lower purities but are not subjected to strong fluid pressures, resulting in higher cell viability of sorted populations. Moreover, only FACS can analyze expression of multiple cell surface markers in a single cell, which can be instrumental in further enrichment of functional BTICs.

Although cell surface markers offer a viable option to isolate and purify BTICs from a tumor cell population, there are caveats associated with it. The cell surface expression of various BTIC markers is in constant flux and evolves rapidly in response to various environmental factors. For example, the cell surface expression of CD133 can be altered by growing cells in hypoxic conditions [75], targeting of glycosylated epitopes [76], and mitochondrial dysfunction induced in long-term culture conditions [77]. Additionally, the use of trypsin for tissue digestion, and the dissociation of neurospheres, into single cells for FACS may also affect cell surface marker and receptor expression on BTICs [78, 79]. Therefore, cell sorting based on surface marker expression offers a snapshot of the biological state of the BTIC rather than elucidating the truly dynamic nature of a tumor's cell population.

References

1. McCulloch EA, Till JE, Siminovitch L (1965) The role of independent and dependent stem cells in the control of hemopoietic and immunologic responses. Wistar Inst Symp Monogr 4:61–68

2. Siminovitch L, McCulloch EA, Till JE (1963) The distribution of colony-forming cells among spleen colonies. J Cell Physiol 62:327–336

3. Till JE, McCulloch EA (1961) A direct measurement of the radiation sensitivity of normal mouse bone marrow cells. Radiat Res 14:213–222

4. Potten CS, Loeffler M (1990) Stem cells: attributes, cycles, spirals, pitfalls and uncertainties. Lessons for and from the crypt. Development 110(4):1001–1020

5. Cattaneo E, McKay R (1990) Proliferation and differentiation of neuronal stem cells regulated by nerve growth factor. Nature 347:762–765

6. Kilpatrick TJ, Bartlett PF (1993) Cloning and growth of multipotential neural precursors: requirements for proliferation and differentiation. Neuron 10:255–265

7. Reynolds BA, Tetzlaff W, Weiss S (1992) A multipotent EGF-responsive striatal embryonic progenitor cell produces neurons and astrocytes. J Neurosci 12:4565–4574

8. Temple S (1989) Division and differentiation of isolated CNS blast cells in microculture. Nature 340:471–473

9. Stemple DL, Anderson DJ (1992) Isolation of a stem cell for neurons and glia from the mammalian neural crest. Cell 71:973–985

10. Lois C, Alvarez-Buylla A (1993) Proliferating subventricular zone cells in the adult mammalian forebrain can differentiate into neurons and glia. Proc Natl Acad Sci U S A 90:2074–2077

11. Reynolds BA, Weiss S (1992) Generation of neurons and astrocytes from isolated cells of the adult mammalian central nervous system. Science 255:1707–1710

12. Gage FH (2000) Mammalian neural stem cells. Science 287:1433–1438

13. McKay R (1997) Stem cells in the central nervous system. Science 276:66–71

14. Rao MS (1999) Multipotent and restricted precursors in the central nervous system. Anat Rec 257:137–148

15. Seaberg RM, van der Kooy D (2002) Adult rodent neurogenic regions: the ventricular subependyma contains neural stem cells, but the dentate gyrus contains restricted progenitors. J Neurosci 22(5):1784–1793

16. Babu H et al (2007) Enriched monolayer precursor cell cultures from micro-dissected adult mouse dentate gyrus yield functional granule cell-like neurons. PLoS One 2(4):e388

17. Uchida N et al (2000) Direct isolation of human central nervous system stem cells. PNAS 97(260):14720–14725

18. Gritti A et al (1995) Basic fibroblast growth factor supports the proliferation of epidermal growth factor-generated neuronal precursor cells of the adult mouse CNS. Neurosci Lett 185(3):151–154

19. Pierce GB, Dixon FJ (1959) Testicular teratomas. I. Demonstration of teratogenesis by metamorphosis of multipotential cells. Cancer 12(3):573–583

20. Illmensee K (1978) Reversion of malignancy and normalized differentiation of teratocarcinoma cells in chimeric mice. Basic Life Sci 12:3–25

21. Mintz B, Illmensee K (1975) Normal genetically mosaic mice produced from malignant teratocarcinoma cells. Proc Natl Acad Sci U S A 72(9):3585–3589

22. Haase D et al (1995) Evidence for malignant transformation in acute myeloid leukemia at the level of early hematopoietic stem cells by cytogenetic analysis of CD34+ subpopulations. Blood 86(8):2906–2912

23. Mehrotra B et al (1995) Cytogenetically aberrant cells in the stem cell compartment (CD34 +lin-) in acute myeloid leukemia. Blood 86 (3):1139–1147

24. Lapidot T et al (1994) A cell initiating human acute myeloid leukaemia after transplantation into SCID mice. Nature 367(6464):645–648

25. Wang JC, Dick JE (2005) Cancer stem cells: lessons from leukemia. Trends Cell Biol 15 (9):494–501

26. Warner JK et al (2004) Concepts of human leukemic development. Oncogene 23 (43):7164–7177

27. Al-Hajj M et al (2003) Prospective identification of tumorigenic breast cancer cells. PNAS 100(7):3983–3988

28. Gotte M, Yip GW (2006) Heparanase, hyaluronan, and CD44 in cancers: a breast carcinoma perspective. Cancer Res 66(21):10233–10237

29. Herrera-Gayol A, Jothy S (1999) Adhesion proteins in the biology of breast cancer: contribution of CD44. Exp Mol Pathol 66 (2):149–156

30. Schabath H et al (2006) CD24 affects CXCR4 function in pre-B lymphocytes and breast carcinoma cells. J Cell Sci 119(Pt 2):314–325

31. Singh SK et al (2003) Identification of a cancer stem cell in human brain tumours. Cancer Res 63:5821–5828

32. Singh SK et al (2004) Identification of human brain tumour initiating cells. Nature 432:396–401

33. Li C et al (2007) Identification of pancreatic cancer stem cells. Cancer Res 67:1030–1037

34. O'Brien CA et al (2007) A human colon cancer cell capable of initiating tumor growth in immunodeficient mice. Nature 445:106–110

35. Ricci-Vitiani L et al (2007) Identification and expansion of human colon-cancer initiating cells. Nature 445:111–115

36. Dalerba P et al (2007) Phenotypic characterization of human colorectal cancer stem cells. Proc Natl Acad Sci USA 104:10158–10163

37. Johnson BE et al (2014) Mutational analysis reveals the origin and therapy-driven evolution

of recurrent glioma. Science 343 (6167):189–193

38. Meyer M et al (2015) Single cell-derived clonal analysis of human glioblastoma links functional and genomic heterogeneity. Proc Natl Acad Sci U S A 112(3):851–856

39. Patel AP et al (2014) Single-cell RNA-seq highlights intratumoral heterogeneity in primary glioblastoma. Science 344 (6190):1396–1401

40. Reinartz R et al (2017) Functional subclone profiling for prediction of treatment-induced Intratumor population shifts and discovery of rational drug combinations in human glioblastoma. Clin Cancer Res 23(2):562–574

41. Sottoriva A et al (2013) Intratumor heterogeneity in human glioblastoma reflects cancer evolutionary dynamics. Proc Natl Acad Sci U S A 110(10):4009–4014

42. Szerlip NJ et al (2012) Intratumoral heterogeneity of receptor tyrosine kinases EGFR and PDGFRA amplification in glioblastoma defines subpopulations with distinct growth factor response. Proc Natl Acad Sci U S A 109 (8):3041–3046

43. Bao S et al (2006) Glioma stem cells promote radioresistance by preferential activation of the DNA damage response. Nature 444 (7120):756–760

44. Liu G et al (2006) Analysis of gene expression and chemoresistance of CD133+ cancer stem cells in glioblastoma. Mol Cancer 5:67

45. Beier D et al (2012) Efficacy of clinically relevant temozolomide dosing schemes in glioblastoma cancer stem cell lines. J Neuro-Oncol 109 (1):45–52

46. Liu J et al (2013) Lung cancer tumorigenicity and drug resistance are maintained through ALDH(hi)CD44(hi) tumor initiating cells. Oncotarget 4(10):1698–1711

47. Ramaswamy V, Taylor MD (2015) The amazing and deadly glioma race. Cancer Cell 28(3):275–277

48. Scorsetti M et al (2015) Multimodality therapy approaches, local and systemic treatment, compared with chemotherapy alone in recurrent glioblastoma. BMC Cancer 15:486

49. Wei W et al (2016) Single-cell phosphoproteomics resolves adaptive signaling dynamics and informs targeted combination therapy in glioblastoma. Cancer Cell 29(4):563–573

50. Kelly PN et al (2007) Tumor growth need not be driven by rare cancer stem cells. Science 317 (5836):337

51. Quintana E et al (2008) Efficient tumor formation by single human melanoma cells. Nature 456:593–598

52. Clarke MF et al (2006) Cancer stem cells—perspectives on current status and future directions: AACR workshop on cancer stem cells. Cancer Res 66:9339–9344

53. Dalerba P, Cho RW, Clarke MF (2007) Cancer stem cells: models and concepts. Annu Rev Med 58:267–284

54. Bonnet D, Dick JE (1997) Human acute myeloid leukemia is organized as a hierarchy that originates from a primitive hematopoietic cell. Nat Med 3(7):730–737

55. Galli R et al (2004) Isolation and characterization of tumorigenic, stem-like neural precursors from human glioblastoma. Cancer Res 64 (19):7011–7021

56. Hemmati HD et al (2003) Cancerous stem cells can arise from pediatric brain tumors. Proc Natl Acad Sci U S A 100 (25):15178–15183

57. Ignatova TN et al (2002) Human cortical glial tumors contain neural stem-like cells expressing astroglial and neuronal markers in vitro. Glia 39(3):193–206

58. Taylor MD et al (2005) Radial glia cells are candidate stem cells of ependymoma. Cancer Cell 8:323–335

59. Beier D et al (2007) CD133(+) and CD133(−) glioblastoma-derived cancer stem cells show differential growth characteristics and molecular profiles. Cancer Res 67:4010–4015

60. Zeppernick F et al (2008) Stem cell marker CD133 affects clinical outcome in glioma patients. Clin Cancer Res 14:123–129

61. Howard BM, Boockvar JA (2008) Stem cell marker CD133 expression predicts outcome in glioma patients. Neurosurgery 62(6):N8

62. Thon N et al (2010) Presence of pluripotent CD133+ cells correlates with malignancy of gliomas. Mol Cell Neurosci 43(1):51–59

63. Son MJ et al (2009) SSEA-1 is an enrichment marker for tumor-initiating cells in human glioblastoma. Cell Stem Cell 4(5):440–452

64. Lathia JD et al (2010) Integrin alpha 6 regulates glioblastoma stem cells. Cell Stem Cell 6 (5):421–432

65. Bao S et al (2008) Targeting cancer stem cells through L1CAM suppresses glioma growth. Cancer Res 68(15):6043–6048

66. Binda E et al (2012) The EphA2 receptor drives self-renewal and tumorigenicity in stem-like tumor-propagating cells from human glioblastomas. Cancer Cell 22 (6):765–780

67. Day BW et al (2013) EphA3 maintains tumorigenicity and is a therapeutic target in glioblastoma multiforme. Cancer Cell 23(2):238–248

68. Nakada M et al (2010) The phosphorylation of ephrin-B2 ligand promotes glioma cell migration and invasion. Int J Cancer 126 (5):1155–1165

69. Alonso MM et al (2011) Genetic and epigenetic modifications of Sox2 contribute to the invasive phenotype of malignant gliomas. PLoS One 6(11):e26740

70. Suva ML et al (2014) Reconstructing and reprogramming the tumor-propagating potential of glioblastoma stem-like cells. Cell 157:580–594

71. Kaneko Y et al (2000) Musashi1: an evolutionally conserved marker for CNS progenitor cells including neural stem cells. Dev Neurosci 22:139–153

72. Venugopal C et al (2012) Bmi1 marks intermediate precursors during differentiation of human brain tumour initiating cells. Stem Cell Res 8(2):141–153

73. Manoranjan B et al (2013) FoxG1 interacts with Bmi1 to regulate self-renewal and tumorigenicity medulloblastoma stem cells. Stem Cells 31(7):1266–1277

74. Fong CY et al (2009) Separation of SSEA-4 and TRA-1-60 labelled undifferentiated human embryonic stem cells from a heterogeneous cell population using magnetic-activated cell sorting (MACS) and fluorescence-activated cell sorting (FACS). Stem Cell Rev 5:72–80

75. Soeda A et al (2009) Hypoxia promotes expansion of the CD133-positive glioma stem cells through activation of HIF-1alpha. Oncogene 28:3949–3959

76. Bidlingmaier S et al (2008) The utility and limitations of glycosylated human CD133 epitopes in defining cancer stem cells. J Mol Med 86:1025–1032

77. Griguer CE et al (2008) CD133 is a marker of bioenergetic stress in human glioma. PLoS One 3:e3655

78. Fukuchi Y et al (2004) Human placenta-derived cells have mesenchymal stem/progenitor cell potential. Stem Cells 22:649–658

79. Schwab KE et al (2008) Identification of surface markers for prospective isolation of human endometrial stromal colony-forming cells. Hum Reprod 23:934–943

Chapter 2

Isolation and Culture of Glioblastoma Brain Tumor Stem Cells

Charles Chesnelong, Ian Restall, and Samuel Weiss

Abstract

Cancer stem cells (CSCs) have been identified in glioblastoma (GBM) and are proposed to be the main actors of post-treatment recurrence contributing to the very dismal prognosis of this devastating disease. Consequently, this important population of cells needs to be further studied to uncover potential vulnerabilities, identify novel therapeutic targets, and develop drugs that can be translated to the clinic. One obstacle preventing progress in understanding the biology of GBM and the development of novel therapies has arguably been the absence of biologically relevant in vitro models representative of the CSC population in GBM. Adherent and non-adherent serum-free culture methods, initially developed for culturing neural stem cells, have been adapted to identify, isolate, maintain, and expand brain tumor stem cells (BTSCs) from GBM. In this chapter, we describe a method to isolate and culture these BTSCs from fresh GBM patient samples.

Key words Neural stem cells, Cancer stem cells, Glioblastoma, Brain tumor stem cell, Neurosphere

1 Introduction

Culture techniques used to isolate and culture brain tumor stem cells (BTSCs) from glioblastoma (GBM) tissue samples [1] originate from our group's seminal study reporting the discovery of neural stem cells (NSCs) in the adult murine brain [2]. This study described the proliferation of NSCs and formation of free-floating clusters of cells termed neurospheres, in serum-free conditions supplemented with epidermal growth factor (EGF). This serum-free approach for isolating and expanding neural precursor cells allowed researchers to study and better understand NSC biology. This NSC culture protocol was later adapted to isolate and culture BTSCs [1, 3–5], making this novel and extremely valuable preclinical model available to glioma researchers. BTSCs fit the cancer stem cell (CSC) definition as they possess the key features of long term self-renewal, multilineage differentiation, and high tumorigenic potential [6]. Importantly, BTSCs are also argued to be more

Sheila K. Singh and Chitra Venugopal (eds.), *Brain Tumor Stem Cells: Methods and Protocols*, Methods in Molecular Biology, vol. 1869, https://doi.org/10.1007/078-1-4939-8805-1_2, © Springer Science+Business Media, LLC, part of Springer Nature 2019

resistant to current therapies [7–9]. The drug resistance, self-renewal, and proliferative capacity characteristic of BTSCs, coupled with their ability to initiate and re-populate tumors when orthotopically xenografted in mice, suggest that these cells are integral to the growth and post-treatment recurrence of GBM (reviewed in [10]). Hence, BTSCs represent a "reservoir of disease" that requires novel therapeutic approaches, if the outcome of GBM is to be improved.

BTSCs retain the genetic and epigenetic alterations of the parental tumors and more closely resemble the tumors of origin than do matching serum-derived cell lines [11]. BTSCs have become an invaluable model to study CSC biology and their role in GBM progression [12] as well as a vital preclinical tool to examine the therapeutic efficacy of various compounds and combinatorial strategies [13, 14]. The development of a large collection of patient-derived BTSC lines provides an unparalleled in vitro model representative of the molecular heterogeneity of GBM [13–15]. Furthermore, BTSC orthotopic xenografts provide an improved in vivo model that recapitulates the diffuse invasion and necrosis patterns of human GBM [14, 15].

2 Materials

2.1 BTSC Media

1. 10× Hormone Mix: Resuspend 50 mg of insulin (Sigma, I5500) in 2 mL of 0.1 N HCl, further dilute the insulin solution with 98 mL of H_2O. Bring to pH 7.4 to dissolve the insulin. As you reach pH 7, insulin will precipitate out of and then go back into solution. Resuspend 200 mg of transferrin (Sigma, T2252) in 20 mL H_2O and add to the insulin solution. Resuspend 19.32 mg of putrescine (Sigma, P7505) in 20 mL H_2O and add to the insulin solution. Add 20 μL of selenium stock solution (stock solution: 1 mg selenium (Sigma, S9133) resuspended in 1.93 mL H_2O). Add 20 μL of progesterone stock solution (stock solution: 1 mg progesterone (Sigma, P6149) resuspended in 1.59 mL EtOH). Top up the hormone mix to 200 mL with H_2O (~40 mL) and filter sterilize. Use this 10× hormone mix fresh to make MHM media or aliquot and freeze at −20 °C.

2. Media Hormone Mix (MHM, *see* **Notes 1–3**): Prepare 10× DMEM/F12 with equal amounts of DMEM (Life Technologies, 12100-046) and F12 (Life Technologies, 21700-075). Add 20 mL of 10× DMEM/F12 to the hormone mix. Add 4 mL of 30% glucose solution (sigma, G7528). Add 3 mL of 7.5% $NaHCO_3$ solution (Sigma, S5761). Add 1 mL of 1 M Hepes (Sigma, H4034). Add 2 mL of 200 mM L-glutamine solution (Sigma, S7513). Filtered sterilize the completed

Table 1
Summary of the final concentrations of principal media components in EF media

Principal components	Final concentration
DMEM/F12	1×
Transferrin	0.1 g/L
Insulin	25 µg/mL
Putrescine	60 µM
Progesterone	0.02 µM
Selenium	0.03 µM
Glucose	9.15 g/L
L-glutamine	4.5 mM
NaHCO$_3$	13.4 mM
Hepes	5 mM
EGF	20 ng/mL
FGF	20 ng/mL
HS	2 µg/mL

MHM media using a bottle top filter (PES, 0.2 µm, Thermo Scientific, 596-4520).

3. EF Media (MHM with growth factors): Measure out required volume of MHM. Add epidermal growth factor (EGF: 20 ng/ mL, Peprotech, recombinant human EGF, AF-100-15). Add fibroblast growth factor (FGF: 20 ng/mL, R&D systems, recombinant human basic FGF, 233-FB). Add heparan sulfate (HS: 2 µg/mL, Sigma, H7640) (*see* Table 1).

2.2 Other Reagents

1. Enzyme solution for tissue digestion: 4 mg collagenase (Sigma, C6885), 4 mg of kynerunic acid (Sigma, K3375), and 10 mg DNase I (Sigma, D4513) resuspended in 20 mL MHM media.

2. Penicillin, streptomycin (Sigma, P4333).

3. Accumax cell dissociation solution (Innovative Cell Technologies, AM105).

2.3 Plasticware and Tissue Culture Equipment

1. Sterile petri dishes.

2. Sterile scalpels.

3. 15 mL polypropylene conical tubes.

4. 50 mL polypropylene conical tubes.

5. Cotton-plugged, sterile serological pipettes.

6. Cell strainers (40 μm nylon mesh, Falcon, 352340).

7. Filter cap tissue culture flasks.

8. Hemocytometer, or another cell counting instrument.

9. Nalgene cryogenic vials.

10. Nalgene freezing container.

11. Pyrex media bottles.

12. Water bath.

13. Aspiration pump.

14. Humidified tissue culture incubator (37 °C, 5% CO_2).

15. Laminar flow hood.

16. Centrifuge.

3   Methods

3.1 Processing GBM Tissue Samples

1. Keep the tissue sample in PBS with antibiotics at 4 °C from collection to processing (*see* **Note 4**).

2. Place sample in a petri dish and select best tumor tissue (*see* **Note 5**).

3. Use a scalpel to finely slice the selected tumor tissue and transfer to a 50 mL polypropylene conical tube.

4. Triturate with a 10 mL pipette until tissue begins to break up and release single cells.

5. If the tissue is not breaking apart and dissociating, an enzyme treatment may be necessary (*see* **Note 6**).

6. Centrifuge the tube at 150 RCF for 5–10 min.

7. Resuspend pellet and pass through a cell strainer.

8. Perform red blood cell lysis (*see* **Note 7**).

9. Centrifuge sample at 150 RCF for 5–10 min.

10. Resuspend in 1–3 mL of EF media depending on the size of the pellet.

11. Count viable cells (*see* **Note 8**).

12. Seed 250,000 to 500,000 cells in a filter cap T25 (7 mL EF media) or 500,000 to 2×10^6 cells in T75 (14 mL EF media) if enough cells are isolated.

3.2 Establishing BTSC Cultures (See Notes 9 and 10)

1. Cells isolated as described above are maintained in 37 °C, 5% CO_2 and fed every week by adding fresh EF media: 2.5 mL to a T25 flask, or 4 mL to a T75 flask. If neurospheres are less than 150 μm in diameter and the media is being depleted or becomes acidic, then half of the media can be replaced (transfer half of the flask into a conical tube/centrifuge/aspirate the

media/add fresh EF media/resuspend/return to flask) (*see* **Note 11**).

2. Cells should be passaged when neurospheres reach 250 μm in diameter, and before the neurosphere centers become dark or the surface of the neurospheres appears to be degrading.

3. Collect media and cells by mixing and washing down the surface of the flask to collect all neurospheres and transfer to a polypropylene conical tube.

4. Centrifuge for 5–10 min at 150 × *g*.

5. Aspirate media, add 1 mL of pre-warmed Accumax solution, and incubate at 37 °C for 5–10 min (minimizing incubation time with Accumax is important for primary cells, *see* **Note 12**).

6. Triturate the Accumax and cell suspension with a P1000 micropipette set at 800 μL 40 times.

7. Add 5 mL of PBS with antibiotics and centrifuge at 150 RCF for 5–10 min.

8. Aspirate the media and resuspend the cells in 1–3 mL of EF media.

9. Count the number of viable cells using an exclusion dye such as trypan blue.

10. Seed cells at ~250,000 cells in a T25 (7 mL EF media), ~500,000 cells in a T75 (14 mL EF media).

<table>
<tr><td>

3.3 Cryogenic Preservation (See Note 13)

</td><td>

1. Collect neurospheres (~150 μm) in a polypropylene conical tube.

2. Centrifuge at 150 RCF for 5 min and aspirate the supernatant.

3. Resuspend neurospheres in MHM media/10% DMSO and transfer into cryogenic vials (*see* **Note 13**).

4. Place cryogenic vials in a freezing container for a minimum of 2 h at −80 °C to allow for gradual freezing.

5. Transfer cryogenic vials to storage at −80 °C.

</td></tr>
<tr><td>

3.4 Recover BTSC Lines from Cryo-Preserved Early Passage Cultures

</td><td>

1. Thaw quickly in 70% ethanol in a beaker in a water bath at 37 °C.

2. Add cells to a 15 mL polypropylene conical tube containing 10 mL of MHM media.

3. Centrifuge the cells (medium sized neurospheres) at 150 × *g* for 5 min.

4. Aspirate all the media and resuspend the cells in 7 mL of EF media into a T25 flask.

5. Passage cells when the neurospheres reach approximately 250 μm in diameter.

6. BTSC passaging and subsequent culture is performed as described in Subheading 3.2, **steps 3–10** (*see* **Note 14**).

</td></tr>
</table>

4   Notes

1. We recommend making media as needed as opposed to stocking large batches. MHM media supplemented with EGF, FGF, and HS is further referred to as EF media. EF media is made fresh and used within 2 weeks.

2. Multiple variants of culture media for BTSCs can be found in the literature. The most commonly used and commercially available version of this media is NeuroCult NS-A Proliferation Kit (Stemcell Technologies, 05751). Different combinations of DMEM/F12, NeuroCult NS-A basal medium (Stemcell Technologies, 05750), or Gibco Neurobasal Medium (Thermo-Fisher Scientific, 21103049) with N-2 Supplement (ThermoFisher Scientific, 17502048) and/or Gibco B-27 Supplement (ThermoFisher Scientific, 12587010) can also be found in the literature. Although not included in this protocol, several studies also report the addition of leukemia inhibitory factor (LIF) to the media [1]. It is important to also note that EGF and FGF (20 ng/mL) are likely in excess in the media and that the EGF and FGF concentration can be decreased to 2–10 ng/mL depending on the cell line with minimal impact on proliferation. In fact, certain BTSC lines have even been reported to not require mitogen supplementation [15].

3. Control over glucose and glutamine concentrations can be important for metabolic studies. Glucose and glutamine concentrations may therefore need to be adjusted. For the best control over these parameters, we recommend preparing media with made to order $10\times$ DMEM/F12 powder without glucose and glutamine (or with any other media components removed, https://www.thermofisher.com/ca/en/home/life-science/bi oproduction/custom-cell-culture-media-and-services/gibco-custom-media-configurator.htmL). Glucose and glutamine can then be easily added to desired concentrations.

4. Tissue samples should be collected in an anonymized manner according to approved institutional research ethics board protocols and procedures that involve proper patient consenting. Following acquisition of the samples from surgery, tissue should be processed immediately. However, the sample can be maintained as intact tissue in MHM media at 4 °C overnight if necessary, in such a case as a late emergency surgery, with minimal impact on viability. In cases where a large amount of good quality tissue is received, one may want to select a piece to be processed for BTSC isolation and process multiple other pieces separately, to be immediately cryogenically preserved for future experiments where primary single cell cryos or pieces of intact tissue will be needed. According to the intended use,

tissue pieces can be snap frozen or formalin fixed and paraffin embedded for future uses including RNA extraction, DNA extraction, protein extraction, IHC, and FISH. Cryopreserved viable single cell cryos can be used in the future to test the original sample composition, such as deep sequencing or single cell RNA sequencing, or to test under specific growth or treatment conditions.

5. A key point for successful isolation of BTSCs is not the quantity as much as the quality of the tissue. Proper selection of the tissue to be processed will increase the chance of deriving a BTSC culture. This will make the process easier by maximizing the tumor to normal tissue ratio, limiting the amount of myelin and other debris that will make it into the culture. The main objective is to avoid normal brain tissue as well as necrotic or bloody tissue and instead selecting good quality tumor tissue. It can be beneficial to perform this selection step in consultation with a pathologist or neurosurgeon. Most good quality GBM tissue can be processed without enzymes by simply scraping tumor cells from the tissue with a scalpel. Following this, light mechanical trituration often achieves sufficient dissociation of cells and ultimately results in cleaner primary cultures.

6. If mechanical trituration is not sufficient to achieve proper dissociation of the tissue, incubate tissue with pre-warmed digestion enzyme solution (*see* Subheading 2.2) and incubate on a shaker at 37 °C for 20 min.

7. RBC lysis is useful to clean up bloody cultures but is not always necessary. Hypotonic lysis is classically used by briefly adding 9 mL of sterile culture grade water to the cell pellet and immediately adjusting osmosis by adding 1 mL of 10× PBS. Ammonium chloride (0.8% solution, Stem Cell Technologies, 07800) can alternatively be added to the cell pellet for 5 min at room temperature, followed by centrifugation and washing. Ammonium chloride appears to not only be more efficient but is also less detrimental to the cells as compared to hypotonic lysis.

8. For cell counting, viable tumor cells should be relatively large and bright under phase contrast. Ignore all debris, possible remaining red blood cells, and smaller, poorly refractive cells, and cells that have not completely excluded trypan blue dye (light blue).

9. For the establishment of BTSC cultures we do not enrich for BTSCs by selecting for or against specific marker(s). CD133 (prominin-1 [16]) has been classically used to identify and enrich patient-derived cultures in radio- or chemo-resistant cells, which more closely resemble NSCs and display enhanced

tumorigenic abilities [5, 9, 17, 18]. However, an increasing number of studies report that CD133 does not identify all GBM stem cell populations [19, 20]. A multitude of other markers have been proposed to enrich for BTSCs [21, 22] but a consensus has yet to be reached to define unified and consistent markers to characterize BTSCs.

10. BTSC primary cells are considered an established BTSC line after five passages in vitro, which can take as little as 2 months or longer than 6 months. One of the major criteria used for BTSC line establishment in our group is that a sufficient number of cells is obtained to make early passage cryos from which cells can be successfully recovered and subsequently passaged.

11. Although cells should grow as bright smooth spheres under phase-contrast microscopy, primary cultures will also contain adherent cells and semi-adherent neurospheres composed of normal (non-transformed cells) or tumor cells that cannot grow as neurospheres. Primary cultures are best kept at high density until they are considered established.

12. BTSCs may display higher or lower tolerance to Accumax. Certain BTSC lines require a shorter incubation time with Accumax to dissociate into single cell suspensions. The incubation time should be minimized to the amount of time needed to effectively dissociate neurospheres in order to maximize cell viability. Some BTSC lines are easily dissociated to single cells by quick and simple mechanical trituration. In this case Accumax may be unnecessary and can be bypassed altogether.

13. When enough cells are obtained, extra flasks should be seeded for cryogenic preservation. BTSCs should be frozen once they grow to medium sized neurospheres of approximately 150 μm in diameter in MHM and 10% DMSO in cryogenic vials. One T25 flask seeded at 250,000 cells should yield enough cells for 3 cryogenic tubes (resuspend pellet in 3 mL MHM—10% DMSO and distribute 1 mL per cryogenic tube). Adding 5% bovine serum albumin (BSA, Sigma, A9647) may increase cell viability upon thawing especially when BTSCs are cryo-preserved as single cells (~1×10^6 cells per cryogenic tube).

14. BTSCs can take a few weeks to properly recover from cryogenic preservation. Following recovery, established BTSC lines can be seeded at a lower density than primary cultures, typically 150,000 in a T25 flask (7 mL EF media) or 300,000 in a T75 flask (14 mL EF media), and is also variable based on the growth rate of different BTSC lines. Typically, BTSC lines need to be passaged every 2–3 weeks, again with variability between lines. BTSCs are typically used within ten passages, after which a new early passage cryo should be thawed again. Importantly BTSCs can also be isolated and maintained

adherently under serum-free conditions (*see* **Note 15**). Regardless of the culture conditions used, it is important to verify the maintenance of the key characteristics of BTSCs: self-renewal (*see* **Note 16**), multi-lineage differentiation potential (*see* **Note 17**), and tumorigenic potential in vivo (*see* **Note 18**).

15. BTSCs are also often grown under adherent conditions that also preserve their CSC properties [23]. For this approach, BTSCs are isolated following a similar protocol to that described here with the distinction that tissue culture plastics are coated with Poly-L-Ornithine (PLO, Sigma, P4957) and laminin (Sigma, P3655). Briefly, coat the bottom of the flask with PLO (20 µg/mL) for 2 h at 37 °C (or overnight at room temperature), remove PLO and coat the flask with laminin (5 µg/mL in PBS) overnight at 37 °C. PLO can be stored at −20 °C (stock solution 0.01% w/v; 0.1 mg/mL), then thawed and further diluted 1:5 in culture grade H_2O for use. Laminin stock solutions of 1 mg/mL can be frozen at −20 °C and should be thawed on ice and diluted to 5 µg/mL in cold PBS. For adherent cultures, to detach and passage adherent BTSCs: remove the media, coat the flask with Accumax for 5 min, add an equal volume of media to the flask, and collect the cell suspension for centrifugation.

16. Maintenance of BTSC self-renewal capacity can be confirmed through serial limiting dilution assays (LDAs) for neurosphere formation in vitro [1] and tumor formation in vivo.

17. To induce differentiation, cells are seeded in MHM alone (growth factor withdrawal) or supplemented with 1% FBS for 7 days (1% FBS promotes differentiation toward the astrocytic and neuronal lineages while growth factor withdrawal allows for oligodendrocytic differentiation) [24]. Alternative differentiation techniques may include treatment with differentiating agents (i.e., BMP4) [23]. To evaluate the ability of BTSCs to differentiate into the three lineages, cells are differentiated on PLO-laminin coated coverslips and expression of markers for the different lineages are then assessed by immunofluorescence (i.e., astrocytic marker—GFAP (Cell Signaling, 3670); neuronal marker—β-Tubulin III (Sigma, T8660); oligodendrocytic marker—O4 (R&D systems, MAB1326).

18. The ability to form tumors as orthotopic xenografts in mice may be the most important characteristic of BTSCs. To evaluate the initiation potential of the BTSCs, in vivo LDAs can be performed where decreasing cell numbers (typically, as low as ten and up to 10,000 cells, [5, 15]) are implanted in the right striatum of 6–8 week old C17/SCID female mice (coordinates, mm from Bregma and dura mater surface: AP +0.5, ML −2.0, DV −3.0). For specific experiments where survival

studies are needed in a reasonable time frame, 50,000–100,000 BTSCs are typically injected. If survival is the endpoint of your specific experiment, mice are typically sacrificed according to animal care guidelines upon significant weight loss or presentation of detrimental neurologic symptoms as per animal care guidelines.

Acknowledgments

This work was supported by a Stem Cell Network grant to S.W.

References

1. Singh SK, Clarke ID, Terasaki M, Bonn VE, Hawkins C, Squire J, Dirks PB (2003) Identification of a cancer stem cell in human brain tumors. Cancer Res 63(18):5821–5828

2. Reynolds BA, Weiss S (1992) Generation of neurons and astrocytes from isolated cells of the adult mammalian central nervous system. Science 255(5052):1707–1710

3. Galli R, Binda E, Orfanelli U, Cipelletti B, Gritti A, De Vitis S, Fiocco R, Foroni C, Dimeco F, Vescovi A (2004) Isolation and characterization of tumorigenic, stem-like neural precursors from human glioblastoma. Cancer Res 64(19):7011–7021. https://doi.org/10.1158/0008-5472.CAN-04-1364

4. Yuan X, Curtin J, Xiong Y, Liu G, Waschsmann-Hogiu S, Farkas DL, Black KL, Yu JS (2004) Isolation of cancer stem cells from adult glioblastoma multiforme. Oncogene 23(58):9392–9400. https://doi.org/10.1038/sj.onc.1208311

5. Singh SK, Hawkins C, Clarke ID, Squire JA, Bayani J, Hide T, Henkelman RM, Cusimano MD, Dirks PB (2004) Identification of human brain tumor initiating cells. Nature 432 (7015):396–401. https://doi.org/10.1038/nature03128

6. Clevers H (2011) The cancer stem cell: premises, promises and challenges. Nat Med 17 (3):313–319. https://doi.org/10.1038/nm.2304

7. Bleau AM, Huse JT, Holland EC (2009) The ABCG2 resistance network of glioblastoma. Cell Cycle 8(18):2936–2944

8. Salmaggi A, Boiardi A, Gelati M, Russo A, Calatozzolo C, Ciusani E, Sciacca FL, Ottolina A, Parati EA, La Porta C, Alessandri G, Marras C, Croci D, De Rossi M (2006) Glioblastoma-derived tumorospheres identify a population of tumor stem-like cells with angiogenic potential and enhanced multidrug resistance phenotype. Glia 54 (8):850–860. https://doi.org/10.1002/glia.20414

9. Bao S, Wu Q, McLendon RE, Hao Y, Shi Q, Hjelmeland AB, Dewhirst MW, Bigner DD, Rich JN (2006) Glioma stem cells promote radioresistance by preferential activation of the DNA damage response. Nature 444 (7120):756–760. https://doi.org/10.1038/nature05236

10. Vescovi AL, Galli R, Reynolds BA (2006) Brain tumor stem cells. Nat Rev Cancer 6 (6):425–436. https://doi.org/10.1038/nrc1889

11. Lee J, Kotliarova S, Kotliarov Y, Li A, Su Q, Donin NM, Pastorino S, Purow BW, Christopher N, Zhang W, Park JK, Fine HA (2006) Tumor stem cells derived from glioblastomas cultured in bFGF and EGF more closely mirror the phenotype and genotype of primary tumors than do serum-cultured cell lines. Cancer Cell 9(5):391–403. https://doi.org/10.1016/j.ccr.2006.03.030

12. Cusulin C, Chesnelong C, Bose P, Bilenky M, Kopciuk K, Chan JA, Cairncross JG, Jones SJ, Marra MA, Luchman HA, Weiss S (2015) Precursor states of brain tumor initiating cell lines are predictive of survival in xenografts and associated with glioblastoma subtypes. Stem Cell Reports 5(1):1–9. https://doi.org/10.1016/j.stemcr.2015.05.010

13. Luchman HA, Stechishin OD, Nguyen SA, Lun XQ, Cairncross JG, Weiss S (2014) Dual mTORC1/2 blockade inhibits glioblastoma brain tumor initiating cells in vitro and in vivo and synergizes with temozolomide to increase orthotopic xenograft survival. Clin Cancer Res 20(22):5756–5767. https://doi.org/10.1158/1078-0432.ccr-13-3389

14. Stechishin OD, Luchman HA, Ruan Y, Blough MD, Nguyen SA, Kelly JJ, Cairncross JG, Weiss S (2013) On-target JAK2/STAT3 inhibition slows disease progression in orthotopic xenografts of human glioblastoma brain tumor stem cells. Neuro-Oncology 15(2):198–207. https://doi.org/10.1093/neuonc/nos302

15. Kelly JJ, Stechishin O, Chojnacki A, Lun X, Sun B, Senger DL, Forsyth P, Auer RN, Dunn JF, Cairncross JG, Parney IF, Weiss S (2009) Proliferation of human glioblastoma stem cells occurs independently of exogenous mitogens. Stem Cells 27(8):1722–1733. https://doi.org/10.1002/stem.98

16. Grosse-Gehling P, Fargeas CA, Dittfeld C, Garbe Y, Alison MR, Corbeil D, Kunz-Schughart LA (2013) CD133 as a biomarker for putative cancer stem cells in solid tumors: limitations, problems and challenges. J Pathol 229(3):355–378. https://doi.org/10.1002/path.4086

17. Lathia JD, Hitomi M, Gallagher J, Gadani SP, Adkins J, Vasanji A, Liu L, Eyler CE, Heddleston JM, Wu Q, Minhas S, Soeda A, Hoeppner DJ, Ravin R, McKay RD, McLendon RE, Corbeil D, Chenn A, Hjelmeland AB, Park DM, Rich JN (2011) Distribution of CD133 reveals glioma stem cells self-renew through symmetric and asymmetric cell divisions. Cell Death Dis 2:e200. https://doi.org/10.1038/cddis.2011.80

18. Brescia P, Ortensi B, Fornasari L, Levi D, Broggi G, Pelicci G (2013) CD133 is essential for glioblastoma stem cell maintenance. Stem Cells 31(5):857–869. https://doi.org/10.1002/stem.1317

19. Zheng X, Shen G, Yang X, Liu W (2007) Most C6 cells are cancer stem cells: evidence from clonal and population analyses. Cancer Res 67(8):3691–3697. https://doi.org/10.1158/0008-5472.CAN-06-3912

20. Beier D, Hau P, Proescholdt M, Lohmeier A, Wischhusen J, Oefner PJ, Aigner L, Brawanski A, Bogdahn U, Beier CP (2007) CD133(+) and CD133(−) glioblastoma-derived cancer stem cells show differential growth characteristics and molecular profiles. Cancer Res 67(9):4010–4015. https://doi.org/10.1158/0008-5472.CAN-06-4180

21. Son MJ, Woolard K, Nam DH, Lee J, Fine HA (2009) SSEA-1 is an enrichment marker for tumor-initiating cells in human glioblastoma. Cell Stem Cell 4(5):440–452. https://doi.org/10.1016/j.stem.2009.03.003

22. Jijiwa M, Demir H, Gupta S, Leung C, Joshi K, Orozco N, Huang T, Yildiz VO, Shibahara I, de Jesus JA, Yong WH, Mischel PS, Fernandez S, Kornblum HI, Nakano I (2011) CD44v6 regulates growth of brain tumor stem cells partially through the AKT-mediated pathway. PLoS One 6(9):e24217. https://doi.org/10.1371/journal.pone.0024217

23. Pollard SM, Yoshikawa K, Clarke ID, Danovi D, Stricker S, Russell R, Bayani J, Head R, Lee M, Bernstein M, Squire JA, Smith A, Dirks P (2009) Glioma stem cell lines expanded in adherent culture have tumor-specific phenotypes and are suitable for chemical and genetic screens. Cell Stem Cell 4(6):568–580. https://doi.org/10.1016/j.stem.2009.03.014

24. Chojnacki A, Weiss S (2008) Production of neurons, astrocytes and oligodendrocytes from mammalian CNS stem cells. Nat Protoc 3(6):935–940. https://doi.org/10.1038/nprot.2008.55

Chapter 3

Establishment and Culture of Patient-Derived Primary Medulloblastoma Cell Lines

Sara Badodi, Silvia Marino, and Loredana Guglielmi

Abstract

Established cell lines have been extensively used in cancer research. They are easy to obtain and expand and are composed of a relatively uniform population of cells. When experimental conditions are kept standard, these cells allow a high reproducibility of experimental findings from independent research groups. However, because these cell lines have been propagated in culture for decades, additional genetic lesions may be acquired leading to modification of their characteristics as compared to the original tumor. Primary cultures represent a valid alternative. Here, we describe standardized protocols to establish medulloblastoma (MB) patient-derived primary cultures from fresh tumor samples. MB primary cells grow as an adherent culture on a laminin coating and can be propagated in vitro for a limited number of passages, therefore reducing the chances to accumulate molecular alterations compared to long-term cultures. Consequently, they better resemble the original tumor both in terms of biological behavior and molecular characteristics. Low-passage MB primary cells can be used as an in vitro model for biochemical studies and functional assays, representing a useful tool to dissect the contribution of molecular pathways to MB pathogenesis. They can also represent a useful screening tool for potential therapeutic agents in preclinical studies.

Key words Medulloblastoma, Cell lines, Primary cultures, Molecular classification, Adherent, Laminin, Epidermal growth factor, Fibroblast growth factor

1 Introduction

Medulloblastoma (MB) is the most common intrinsic malignant tumor affecting the central nervous system in childhood. Genetic, epigenetic, and transcriptomic studies have subclassified MBs in four distinct molecular subgroups (WNT, SHH, Group 3, and Group 4) with different prognosis and response to therapy [1, 2].

WNT MBs are characterized by altered WNT signaling frequently linked to mutations in CTNNB1, leading to increased nuclear localization of β-catenin [3, 4]. SHH subgroup shows deregulation of SHH signaling due to mutations in any component of the pathway (i.e., PTCH1, SMO, SUFU, GLI1, and GLI2) [5, 6]. G3 and G4 groups remain less well characterized and,

Sheila K. Singh and Chitra Venugopal (eds.), *Brain Tumor Stem Cells: Methods and Protocols*, Methods in Molecular Biology, vol. 1869, https://doi.org/10.1007/978-1-4939-8805-1_3, © Springer Science+Business Media, LLC, part of Springer Nature 2019

although they share some chromosomal abnormalities and the altered expression of certain genes (e.g., OTX2), they are molecularly distinct [1]. G3 group is usually referred to as the MYC subgroup even though only 17% of G3 patients exhibit MYC amplifications [7]. G4 MBs harbor frequent amplification of MYCN and cyclin-dependent kinase 6 (CDK6) [2] and LMX1A, LHX2, and EOMES have been proposed as major regulators of this subgroup [8]. Recently, the heterogeneity within these subgroups has been reduced by two independent studies which have defined additional subtypes [9, 10].

The current therapy for MB entails complete surgical resection, whenever possible, followed by adjuvant chemotherapy and craniospinal irradiation. This multimodal treatment does not take into account the molecular subgroups of MB and although it has increased survival up to 80% [11], it also induces severe treatment-related morbidity and an impaired quality of life in many survivors. A major challenge of MB treatment is to identify which molecular subgroup mostly benefits from the different chemotherapy agents currently used and which new potential drug could be more effective for the specific subgroups and subtypes. In this scenario, the availability of in vitro models to screen for novel therapeutic approaches in preclinical studies becomes of paramount importance.

In the past four decades, 44 continuous MB cell lines have been established and only less than half of these has been classified according to the current molecular characterization [12]. The majority of MB cell lines belongs to G3 harboring MYC amplification and to SHH group with mutations in TP53. Only few established cell lines represent the WNT and G4 subtypes, despite the high frequency of these subgroups in patients. Therefore, current in vitro models are certainly underrepresenting the MB heterogeneity. Moreover, it is widely accepted that the extensive propagation of cell cultures in vitro may lead to severe genetic modifications and changes in their molecular characteristics leading to alteration of their biological behavior.

Developing models that retain as much as possible the features of the original tumors including their molecular classification, is a major aim for researchers to enhance the clinical relevance of their results. Primary cultures of patient-derived tumors have the potential of better mirroring the original features of the patients affected by the various subgroups of MB when maintained in cultures for a limited number of passages. Alternatively, patient-derived MB cells can be cultured and propagated in immunocompromised mice through orthotopic xenografts with the advantages that in this setting the cells can grow in a more physiologic microenvironment [13, 14].

Here, we present a standardized protocol to obtain patient-derived MB primary cultures from fresh primary tumors. Tumor tissue is collected during tumor debulking and the fresh specimen is transferred from the OP Theatre to the pathology laboratory for tissue diagnosis. If an intraoperative assessment is performed, guidance can be obtained on which part of the specimen contains neoplastic tissue. Surplus tumor tissue, which is not required for definite tissue diagnosis, is processed to derive primary MB cultures, if an appropriate consent to the research and/or cell banking is available. After tumor sample dissociation, the cells are maintained as adherent monolayer cultures in a serum-free medium supplemented with Epidermal Growth Factor (EGF) and Fibroblast Growth Factor (FGF). We and others have applied these culturing conditions to MB and other brain tumor-derived cell cultures and showed good consistency in terms of (1) molecular similarity with the original tumor and (2) enrichment for a population of tumor-initiating cells within the culture [15–17]. Low-passage MB primary cells have reduced chances to accumulate molecular alteration as compared to long-term cultures and they can be compared with the original tumor to assess their similarity in a range of different genome-wide approaches (e.g., methylation profile, transcriptomic analysis). Hence, they represent a useful in vitro model for cell-based assays to study altered molecular pathways leading to MB formation and screen for new therapeutic agents.

2   Materials

2.1   Reagents

1. NeuroCult NS-A proliferation kit (STEMCELL Technologies).
2. Penicillin/Streptomycin.
3. Epidermal Growth Factor (EGF), recombinant mouse, 100 μg/mL stock in sterile dH_2O (*see* **Note 1**).
4. Basic Fibroblast Growth Factor (bFGF), recombinant human, 100 μg/mL stock in sterile PBS (*see* **Note 1**).
5. Heparin (STEMCELL Technologies, 2 mg/mL).
6. Laminin (Sigma, L2020, 1 mg/mL).
7. PBS with Ca+ and Mg+.
8. ACK Lysing Buffer.
9. Accutase.
10. STEM-CELLBANKER (Ambsio, ZENOAQ).
11. Cell strainer 70 μm.
12. 10 cm petri dishes.
13. Costar® 6-Well TC-Treated Plates (Corning).

2.2 Laminin Coating

1. Thaw laminin gradually at 4 °C.

2. Dilute laminin at a final concentration of 10 µg/mL in PBS and add to the plate (*see* **Note 2**).

3. Incubate the plate for 3 h at 37 °C in the incubator.

4. Wash the plate twice with PBS and remove PBS just before plating. Plate must not dry out.

2.3 NeuroCult NS-A Proliferation Culture Medium Preparation

1. Thaw NeuroCult NS-A Proliferation Supplement (Human, 50 mL) overnight at 4 °C or quickly at 37 °C and then add to NeuroCult NS-A Basal Medium (Human, 450 mL).

2. Add 10 mL Penicillin/Streptomycin 50X and prepare aliquots of 45 mL of complete medium to store at −20 °C (*see* **Note 3**).

3. Freshly add EGF at a final concentration of 20 ng/µL (dilution 1:5000), bFGF at a final concentration of 10 ng/µL (dilution 1:10,000), and Heparin at a final concentration of 2 µg/µL (dilution 1:1000) to obtain culture medium (*see* **Note 4**).

3 Methods

This protocol describes the preparation of primary cultures of MB. However, it can be used for the derivation of primary cultures from other types of brain tumors, such as high-grade and low-grade gliomas.

3.1 Establishing Primary Cultures of MB from Tumor Specimen

The entire procedure must be carried out in a tissue culture laboratory equipped with a sterile hood.

1. Prepare working area spraying surfaces with 70% ethanol.

2. Place the tissue in a 10 cm petri dish (*see* **Note 5**) and mince the tissue as finely as possible using a one-sided razor blade and sterile forceps (*see* **Note 6**). Transfer the tissue into a 50 mL tube using PBS.

3. Wash the dish with PBS to collect as much as possible of the minced tissue and transfer in the same 50 mL tube.

4. Centrifuge at room temperature for 5 min at 200 × g and discard the supernatant.

5. Incubate the tissue in ACK buffer for 20 min at room temperature to lyse red blood cells (*see* **Note 7**). Tap the tube every 5 min to make sure the ACK buffer properly permeates the tissue.

6. Centrifuge at room temperature for 5 min at 200 × g, discard the supernatant, and wash with PBS.

7. Centrifuge for 5 min at 200 × g and discard the supernatant.

8. To enzymatically dissociate the tissue, resuspend the pellet in 5 mL of Accutase solution and incubate for 10 min at 37 °C.

9. Stop the reaction by adding an equal volume of freshly prepared complete medium.

10. Centrifuge for 5 min at 200 × g and discard the supernatant.

11. Add 1 mL of freshly prepared culture medium and pipette up and down few times to resuspend the cells and enhance mechanical dissociation of the tissue.

12. Top up to the desired final volume with freshly prepared culture medium and resuspend again (*see* **Note 8**).

13. Filter using a 70 μm cell strainer to remove parts of tissue which are not completely dissociated.

14. Seed the cells in a 6-well laminin-coated plate (*see* **Note 9**).

3.2 Maintaining Primary Cultures of MB

1. The culture must be checked daily (*see* Fig. 1a–f for an example of culture just after the dissociation and after several passages) and cells should be washed to remove debris, if present. As most of the cells will die within few days, the pH of the culture medium could dramatically drop preventing the establishment of the culture.

2. While the culture is being established, cells with the ability to propagate in vitro will release soluble factors in the culture medium, necessary for the success of the procedure. To avoid the withdrawal of these factors it is important to perform an half media change and not a complete media change every other day. After about 1 week, it should become clear if the culture was successfully established or not.

3.3 Passaging

Cells should be passaged when they reach 70%–90% confluency. The protocol described below is optimized for cells grown in a 6-well (9 cm^2) plate. If using plates of a larger or smaller surface scale up or down volumes of all the reagents.

1. Remove the culture medium and wash the cells with 1 mL of PBS.

2. Add 1 mL of Accutase solution to the cells and place in the incubator at 37 °C for a maximum of 5 min.

3. Check at the microscope to confirm that cells are completely detached; if not gently tap the flask to facilitate detachment.

4. Block reaction with 1 mL of complete medium.

5. Centrifuge at room temperature for 5 min at 200 × g.

6. Discard the supernatant and resuspend the cells in a freshly prepared culture medium (*see* **Note 10**).

7. Aspirate PBS from laminin-coated plate and seed cells.

8. Check the cells at the microscope to confirm they are evenly plated and no aggregates or clumps are present (*see* **Note 11**).

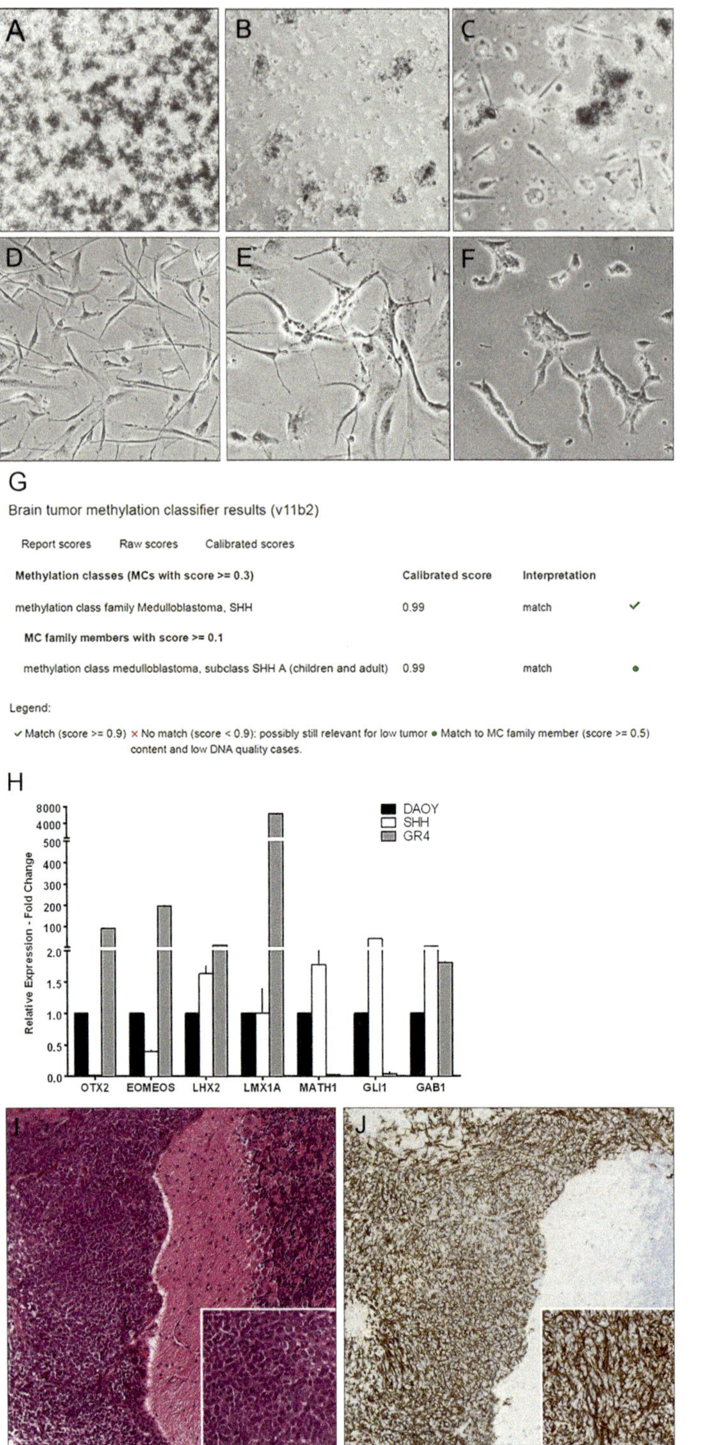

Fig. 1 (**a–f**) Establishment of a primary MB culture from surgical tissue. (**a–c**) Isolation of MB primary cells. Appearance of MB culture immediately after specimen dissociation (**a**) showing mixed population of different brain cell

3.4 Cryopreserving

If the culture is successful, it is highly recommended to start cryopreservation in order to bank cells at early passages.

1. Collect cell pellet from 80–90% confluent well of a 6-well plate as described above (Subheading 3.3). Resuspend the pellet in 1 mL of Stem Cell-Banker and transfer the cells into a labeled cryovial and then into Mr. Frosty container (*see* **Note 12**).

2. To recover cells from cryopreservation, rapidly thaw a cryovial at 37 °C in a waterbath. Transfer the cell suspension in a pre-warmed culture medium and centrifuge at room temperature for 5 min at $200 \times g$. Resuspend pellet in a freshly prepared culture medium and seed in a laminin-coated well of a 6-well plate. After an overnight incubation, when the cells are attached, remove the medium and replace with a new freshly prepared culture medium.

3.5 Subgroup Characterization

Multiple approaches can be used to characterize the molecular subgroup to which the primary MB culture belongs. The gold standard is the analysis of the methylation profile of the cells [18, 19]. A targeted expression analysis of genes known to be deregulated specifically in the different subgroups can also be used. DNA and RNA for these analyses can be obtained from the same cell pellet using commercially available kits (*see* **Note 13**).

1. Perform bisulphite conversion on 200–500 ng of DNA with EZ DNA Methylation™ Kit and process the DNA on Illumina EPIC 850 K. Data obtained can be analyzed with classifier https://www.molecularneuropathology.org/mnp [20] (Fig. 1g).

2. Reverse transcribe 500 ng of RNA using Superscript III reverse transcriptase (Invitrogen). Perform qPCR on 10 ng of the obtained cDNA with TaqMan® Gene Expression Master Mix (Applied Biosystem) and specific gene expression assays (Table 1). *See* Fig. 1h for an example of expression analysis.

Fig. 1 (continued) types and tissue clumps. During first days of culture (**b**) debris and dead cells are removed while viable cells start to adhere. After about a week (**c**) the culture is successfully established and adherent cells begin to proliferate. (**d–f**): Example of MB primary cultures derived from independent patients. Phase-contrast images (**a, b**: 10×, **c–f**: 20×). (**g**) Example of MB subgroup classification based on methylation profile. (**h**) Expression of genes involved in MB subgroups (OTX2:G3/4, EOMEOS and LMX1A: G4, MATH1 and GLI1: SHH) in two independent primary cultures and a SHH cell line (DAOY). Culture A shows an expression profile related to SHH group while culture B to G4. To be noticed that primary cultures derived from SHH group MB express higher levels of MATH1 and GLI1 as compared to DAOY, highlighting how cell lines can be substantially different in terms of gene expression from the tumor they were originally isolated from. H&E (**i**) shows morphological features of MB and human Vimentin staining (**j**) confirms the human origin of the tumor in a xenograft model (10×, insets, 40×)

Table 1
Applied biosystems TaqMan assay ID for target genes used in this study

Target gene	TaqMan assay ID	MB subtype
OTX2	Hs00222238_m1	G3 [21]–G4 [1]
EOMES	Hs00172872_m1	G4 [8]
LMX1A	Hs00892663_m1	G4 [8]
MATH1 (ATOH1)	Hs00245453_s1	SHH [22]
GLI1	Hs00171790_m1	SHH [5, 6]

3.6 Lentiviral Transduction

MB primary cells can be transduced with lentiviral vectors to induce genetic modifications. The protocol described below must be optimized for each primary cell line obtained in order to achieve a good percentage of cell transduction without induction of overt cell death or differentiation.

1. Seed $0.25–1 \times 10^6$ cells in a well of a 6-well plate. Choose your cell density to reach around 60% of confluence upon transduction.

2. Infect the cells at a multiplicity of infection (MOI) of 2–12 and let the cells incubate with the viral particles for 16–20 h at 37 °C in the incubator (*see* **Note 14**).

3. Remove the medium with lentiviral particles and replace with a fresh culture medium.

4. If the viral vector contains an antibiotic selection, select the infected cells using fresh culture medium with an adequate concentration of this antibiotic (*see* **Note 15**) in order to obtain resistant colonies that have stably integrated the vector.

3.7 Neurosphere Formation Assay

The identification of self-renewing and multipotent neural stem cells (NSCs) in mammalian brain has been a major breakthrough for modeling different neurological diseases, including brain tumors. When NSCs were first isolated by Reynolds and Weiss [23] they were cultured in serum-free conditions and in the presence of mitogens factors. Cells isolated from the striatum of adult mouse brain were seeded at low density, in non-adherent conditions and a small proportion (~1.5%) of these cells underwent cell division in vitro. This assay allows the culturing of NCSs as three-dimensional clusters, known as neurospheres, which resemble the typical structure and composition of the NSC niche. Neurospheres are therefore very heterogeneous, harboring only 2.5% of true NSCs within a sphere [24]. However, thanks to their ability to reproduce to some degree the complexity of the NSC niche in vitro, they represent a valid tool to preserve the cellular heterogeneity and investigate the capability of cells isolated from primary tumors such as MB to generate spheroid structures [25].

It is important to maintain the cells at low density when performing the neurosphere formation assay to assess the clonogenity of the sphere.

1. Seed the cells into low-attachment 96-well plates at a density of 5–10 cells/µL in 100 µL (between 150 and 300 cells/cm^2).

2. Perform a serial dilution to obtain the lowest density possible to generate neurospheres from clonal single cells (*see* **Note 16**).

3. Change the medium every other day and over the course of one week neurospheres should be visible (*see* **Note 17**).

3.8 In Vivo Orthotopic Xenografts

Orthotopic injection of the cells into immunodeficient mice is the ultimate approach to test the oncogenic potential of the MB primary culture obtained from patient tumor. Furthermore, it is possible to screen whether genetic modification (e.g., ectopic expression or knockdown of genes) or drug treatment of the cells can affect their tumor formation capacity.

1. Resuspend 5–10 × 10^5 cells in 2 µL of sterile PBS and inject them in the right cerebellar hemisphere (2 mm lateral and 2 mm posterior to lambda, 2 mm deep) with a 26-gauge Hamilton syringe needle. Cull mice when they show neurological signs, collect cerebellum and brain stem, and fix 10% neutral-buffered formalin. Tissue can be either cryopreserved in OCT or fixed in formalin and embedded in paraffin for further histological and immunohistochemical analysis.

2. To confirm the origin of the tumor, an immunostaining for human vimentin can be carried out. Furthermore, immunostaining for synaptophysin or MAP 2 can be performed to confirm the morphological diagnosis (Fig. 1i and j).

---

4 Notes

1. EGF and FGF stock aliquots must be stored at −20 °C. Avoid multiple thawing and freezing.

2. The volume of laminin should cover the entire area of the flask/plate that you are using. Make sure that the laminin does not dry out during the coating process.

3. Complete medium is stable for up to 2 weeks at 4 °C.

4. Do not store and reuse culture medium after EGF, bFGF, and Heparin have been added. EGF and bFGF are particularly unstable at 4 °C, thus they require to be added freshly before media change or subculturing.

5. Try to process the tissue as soon as possible to prevent excessive cell death.

6. During the mincing procedure, it is very important that tissue does not dry out. Add either PBS 1× or complete medium during this step.

7. The amount of ACK buffer should be enough to cover the tissue.

8. The amount of freshly prepared culture medium to add is dependent on the quantity of tissue. As a general guide, use 3 mL of culture medium per well for a 6-well plate.

9. It is difficult to determine the number of cells at the beginning of the culture as the sample contains debris and clumps (*see* Fig. 1a). A good starting point to establish a new culture is to seed cells in 1 or 2 wells of a 6-well plate. The majority of the cells will indeed die and only a small proportion of them will be capable to proliferate in vitro, thus excessive dilution might result in a not sufficient number of cells to allow the culture to be established.

10. Cells originating from different biopsies might present different doubling times in vitro. However, cells must not be excessively diluted. Subculturing in the range of 1:2 or 1:3 maximum is ideal for most of the specimen.

11. First resuspend the cells in a small amount of freshly prepared culture medium, for example 1 mL. Use the P1000 pipette to resuspend the cells and obtain a homogenous single cell suspension. After that, top up to final volume and resuspend the cells with a 5 mL pipette. If aggregates and clumps are still present, use 1000 pipette to dissociate them.

12. Mr. Frosty must be left overnight at −80 °C. Cells can be transferred to liquid nitrogen for long-term storage the next day. It is not recommended to keep the cells at −80 °C for long-term as the recovery rate sensibly drops after few weeks. It is also important that Mr. Frosty is at room temperature prior to use and isopropanol is added fresh every three usages to ensure slow freezing of the cells.

13. We extract DNA and RNA with the RNA/DNA/Protein Purification Plus Kit (Norgen Biotek) and we routinely use these for high-throughput sequencing.

14. MOI and hours of incubation with the virus should be optimized for each primary culture obtained in order to achieve and optimal transduction efficiency.

15. The concentration of antibiotics needed to select transduced cells varies for each cell type. Hence, a titration experiment is highly recommended for each new primary line obtained.

(a) Seed $0.8\text{–}3.0 \times 10^5$ cells per well in a 24-well plate, choose a cell density that permits reaching 60–80% confluency upon antibiotic addition.

(b) Add antibiotic in increasing concentration to each well, make sure to have an untreated well.

(c) Typical ranges of concentrations are: Puromycin (0.25–10 μg/mL), Hygromycin B (100–500 μg/mL), and G418 (0.1–2.0 mg/mL).

(d) Every 2 days replace the medium with freshly added antibiotic for up to a week. Check the cells every day to determine the lowest concentration necessary to kill all the cells.

(e) Use the concentration obtained from the antibiotic titration to select the transduced cells.

16. Each line might behave differently, however performing a 1:2 serial dilution should result in a series of low densities plating useful to assess the ability of single cells to clonally generate new spheroid structures. Ten serial dilutions are generally enough for this test.

17. Ideally, each well should contain one single sphere. If a serial dilution was carried out it should be possible to observe how the lowest plating densities generally produce one single sphere per well while higher plating densities generate multiple spheroid structures. This indicates that the cells were not seeded at clonal density and aggregates, rather than spheres, may have formed.

References

1. Northcott PA, Korshunov A, Witt H, Hielscher T, Eberhart CG, Mack S, Bouffet E, Clifford SC, Hawkins CE, French P, Rutka JT, Pfister S, Taylor MD (2011) Medulloblastoma comprises four distinct molecular variants. J Clin Oncol 29(11):1408–1414. https://doi.org/10.1200/JCO.2009.27.4324

2. Taylor MD, Northcott PA, Korshunov A, Remke M, Cho YJ, Clifford SC, Eberhart CG, Parsons DW, Rutkowski S, Gajjar A, Ellison DW, Lichter P, Gilbertson RJ, Pomeroy SL, Kool M, Pfister SM (2012) Molecular subgroups of medulloblastoma: the current consensus. Acta Neuropathol 123(4):465–472. https://doi.org/10.1007/s00401-011-0922-z

3. Thompson MC, Fuller C, Hogg TL, Dalton J, Finkelstein D, Lau CC, Chintagumpala M, Adesina A, Ashley DM, Kellie SJ, Taylor MD, Curran T, Gajjar A, Gilbertson RJ (2006) Genomics identifies medulloblastoma subgroups that are enriched for specific genetic alterations. J Clin Oncol 24(12):1924–1931. https://doi.org/10.1200/JCO.2005.04.4974

4. Clifford SC, Lusher ME, Lindsey JC, Langdon JA, Gilbertson RJ, Straughton D, Ellison DW (2006) Wnt/wingless pathway activation and chromosome 6 loss characterize a distinct molecular sub-group of medulloblastomas associated with a favorable prognosis. Cell Cycle 5(22):2666–2670. https://doi.org/10.4161/cc.5.22.3446

5. Taylor MD, Liu L, Raffel C, Hui CC, Mainprize TG, Zhang X, Agatep R, Chiappa S, Gao L, Lowrance A, Hao A, Goldstein AM, Stavrou T, Scherer SW, Dura WT, Wainwright B, Squire JA, Rutka JT, Hogg D (2002) Mutations in SUFU predispose to medulloblastoma. Nat Genet 31(3):306–310. https://doi.org/10.1038/ng916

6. Kool M, Jones DT, Jager N, Northcott PA, Pugh TJ, Hovestadt V, Piro RM, Esparza LA,

Markant SL, Remke M, Milde T, Bourdeaut F, Ryzhova M, Sturm D, Pfaff E, Stark S, Hutter S, Seker-Cin H, Johann P, Bender S, Schmidt C, Rausch T, Shih D, Reimand J, Sieber L, Wittmann A, Linke L, Witt H, Weber UD, Zapatka M, Konig R, Beroukhim R, Bergthold G, van Sluis P, Volckmann R, Koster J, Versteeg R, Schmidt S, Wolf S, Lawerenz C, Bartholomae CC, von Kalle C, Unterberg A, Herold-Mende C, Hofer S, Kulozik AE, von Deimling A, Scheurlen W, Felsberg J, Reifenberger G, Hasselblatt M, Crawford JR, Grant GA, Jabado N, Perry A, Cowdrey C, Croul S, Zadeh G, Korbel JO, Doz F, Delattre O, Bader GD, McCabe MG, Collins VP, Kieran MW, Cho YJ, Pomeroy SL, Witt O, Brors B, Taylor MD, Schuller U, Korshunov A, Eils R, Wechsler-Reya RJ, Lichter P, Pfister SM, Project IPT (2014) Genome sequencing of SHH medulloblastoma predicts genotype-related response to smoothened inhibition. Cancer Cell 25(3):393–405. https://doi.org/10.1016/j.ccr.2014.02.004

7. Roussel MF, Robinson GW (2013) Role of MYC in Medulloblastoma. Cold Spring Harb Perspect Med 3(11). https://doi.org/10.1101/cshperspect.a014308

8. Lin CY, Erkek S, Tong Y, Yin L, Federation AJ, Zapatka M, Haldipur P, Kawauchi D, Risch T, Warnatz HJ, Worst BC, Ju B, Orr BA, Zeid R, Polaski DR, Segura-Wang M, Waszak SM, Jones DT, Kool M, Hovestadt V, Buchhalter I, Sieber L, Johann P, Chavez L, Groschel S, Ryzhova M, Korshunov A, Chen W, Chizhikov VV, Millen KJ, Amstislavskiy V, Lehrach H, Yaspo ML, Eils R, Lichter P, Korbel JO, Pfister SM, Bradner JE, Northcott PA (2016) Active medulloblastoma enhancers reveal subgroup-specific cellular origins. Nature 530(7588):57–62. https://doi.org/10.1038/nature16546

9. Cavalli FMG, Remke M, Rampasek L, Peacock J, Shih DJH, Luu B, Garzia L, Torchia J, Nor C, Morrissy AS, Agnihotri S, Thompson YY, Kuzan-Fischer CM, Farooq H, Isaev K, Daniels C, Cho B-K, Kim S-K, Wang K-C, Lee JY, Grajkowska WA, Perek-Polnik M, Vasiljevic A, Faure-Conter C, Jouvet A, Giannini C, Nageswara Rao AA, Li KKW, Ng H-K, Eberhart CG, Pollack IF, Hamilton RL, Gillespie GY, Olson JM, Leary S, Weiss WA, Lach B, Chambless LB, Thompson RC, Cooper MK, Vibhakar R, Hauser P, van Veelen M-LC, Kros JM, French PJ, Ra YS, Kumabe T, López-Aguilar E, Zitterbart K, Sterba J, Finocchiaro G, Massimino M, Van Meir EG, Osuka S, Shofuda T, Klekner A, Zollo M, Leonard JR, Rubin JB, Jabado N, Albrecht S, Mora J, Van Meter TE, Jung S, Moore AS, Hallahan AR, Chan JA, Tirapelli DPC, Carlotti CG, Fouladi M, Pimentel J, Faria CC, Saad AG, Massimi L, Liau LM, Wheeler H, Nakamura H, Elbabaa SK, Perezpeña-Diazconti M, Chico Ponce de León F, Robinson S, Zapotocky M, Lassaletta A, Huang A, Hawkins CE, Tabori U, Bouffet E, Bartels U, Dirks PB, Rutka JT, Bader GD, Reimand J, Goldenberg A, Ramaswamy V, Taylor MD (2017) Intertumoral heterogeneity within medulloblastoma subgroups. Cancer Cell 31(6):737–754.e6

10. Schwalbe EC, Lindsey JC, Nakjang S, Crosier S, Smith AJ, Hicks D, Rafiee G, Hill RM, Iliasova A, Stone T, Pizer B, Michalski A, Joshi A, Wharton SB, Jacques TS, Bailey S, Williamson D, Clifford SC (2017) Novel molecular subgroups for clinical classification and outcome prediction in childhood medulloblastoma: a cohort study. Lancet Oncol 18(7):958–971

11. Gottardo NG, Hansford JR, McGlade JP, Alvaro F, Ashley DM, Bailey S, Baker DL, Bourdeaut F, Cho YJ, Clay M, Clifford SC, Cohn RJ, Cole CH, Dallas PB, Downie P, Doz F, Ellison DW, Endersby R, Fisher PG, Hassall T, Heath JA, Hii HL, Jones DT, Junckerstorff R, Kellie S, Kool M, Kotecha RS, Lichter P, Laughton SJ, Lee S, McCowage G, Northcott PA, Olson JM, Packer RJ, Pfister SM, Pietsch T, Pizer B, Pomeroy SL, Remke M, Robinson GW, Rutkowski S, Schoep T, Shelat AA, Stewart CF, Sullivan M, Taylor MD, Wainwright B, Walwyn T, Weiss WA, Williamson D, Gajjar A (2014) Medulloblastoma down under 2013: a report from the third annual meeting of the international Medulloblastoma working group. Acta Neuropathol 127(2):189–201. https://doi.org/10.1007/s00401-013-1213-7

12. Ivanov DP, Coyle B, Walker DA, Grabowska AM (2016) In vitro models of medulloblastoma: choosing the right tool for the job. J Biotechnol 236:10–25. https://doi.org/10.1016/j.jbiotec.2016.07.028

13. Shu Q, Wong KK, Su JM, Adesina AM, Yu LT, Tsang YT, Antalffy BC, Baxter P, Perlaky L, Yang J, Dauser RC, Chintagumpala M, Blaney SM, Lau CC, Li XN (2008) Direct orthotopic transplantation of fresh surgical specimen preserves CD133+ tumor cells in clinically relevant mouse models of medulloblastoma and glioma. Stem Cells 26(6):1414–1424. https://doi.org/10.1634/stemcells.2007-1009

14. Zhao X, Liu Z, Yu L, Zhang Y, Baxter P, Voicu H, Gurusiddappa S, Luan J, Su JM,

Leung HC, Li XN (2012) Global gene expression profiling confirms the molecular fidelity of primary tumor-based orthotopic xenograft mouse models of medulloblastoma. Neuro-Oncology 14(5):574–583. https://doi.org/10.1093/neuonc/nos061

15. Lee J, Kotliarova S, Kotliarov Y, Li A, Su Q, Donin NM, Pastorino S, Purow BW, Christopher N, Zhang W, Park JK, Fine HA (2006) Tumor stem cells derived from glioblastomas cultured in bFGF and EGF more closely mirror the phenotype and genotype of primary tumors than do serum-cultured cell lines. Cancer Cell 9(5):391–403. https://doi.org/10.1016/j.ccr.2006.03.030

16. Clavreul A, Jean I, Preisser L, Chassevent A, Sapin A, Michalak S, Menei P (2009) Human glioma cell culture: two FCS-free media could be recommended for clinical use in immunotherapy. In Vitro Cell Dev Biol Anim 45(9):500–511. https://doi.org/10.1007/s11626-009-9215-4

17. Sanden E, Eberstal S, Visse E, Siesjo P, Darabi A (2015) A standardized and reproducible protocol for serum-free monolayer culturing of primary paediatric brain tumors to be utilized for therapeutic assays. Sci Rep 5:12218. https://doi.org/10.1038/srep12218

18. Hovestadt V, Jones DT, Picelli S, Wang W, Kool M, Northcott PA, Sultan M, Stachurski K, Ryzhova M, Warnatz HJ, Ralser M, Brun S, Bunt J, Jager N, Kleinheinz K, Erkek S, Weber UD, Bartholomae CC, von Kalle C, Lawerenz C, Eils J, Koster J, Versteeg R, Milde T, Witt O, Schmidt S, Wolf S, Pietsch T, Rutkowski S, Scheurlen W, Taylor MD, Brors B, Felsberg J, Reifenberger G, Borkhardt A, Lehrach H, Wechsler-Reya RJ, Eils R, Yaspo ML, Landgraf P, Korshunov A, Zapatka M, Radlwimmer B, Pfister SM, Lichter P (2014) Decoding the regulatory landscape of medulloblastoma using DNA methylation sequencing. Nature 510(7506):537–541. https://doi.org/10.1038/nature13268

19. Schwalbe EC, Williamson D, Lindsey JC, Hamilton D, Ryan SL, Megahed H, Garami M, Hauser P, Dembowska-Baginska B, Perek D, Northcott PA, Taylor MD, Taylor RE, Ellison DW, Bailey S, Clifford SC (2013) DNA methylation profiling of medulloblastoma allows robust subclassification and improved outcome prediction using formalin-fixed biopsies. Acta Neuropathol 125(3):359–371. https://doi.org/10.1007/s00401-012-1077-2

20. Capper D, Jones DTW, Sill M, Hovestadt V, Schrimpf D, Sturm D, Koelsche C, Sahm F, Chavez L, Reuss DE, Kratz A, Wefers AK, Huang K, Pajtler KW, Schweizer L, Stichel D, Olar A, Engel NW, Lindenberg K, Harter PN, Braczynski AK, Plate KH, Dohmen H, Garvalov BK, Coras R, Hölsken A, Hewer E, Bewerunge-Hudler M, Schick M, Fischer R, Beschorner R, Schittenhelm J, Staszewski O, Wani K, Varlet P, Pages M, Temming P, Lohmann D, Selt F, Witt H, Milde T, Witt O, Aronica E, Giangaspero F, Rushing E, Scheurlen W, Geisenberger C, Rodriguez FJ, Becker A, Preusser M, Haberler C, Bjerkvig R, Cryan J, Farrell M, Deckert M, Hench J, Frank S, Serrano J, Kannan K, Tsirigos A, Brück W, Hofer S, Brehmer S, Seiz-Rosenhagen M, Hänggi D, Hans V, Rozsnoki S, Hansford JR, Kohlhof P, Kristensen BW, Lechner M, Lopes B, Mawrin C, Ketter R, Kulozik A, Khatib Z, Heppner F, Koch A, Jouvet A, Keohane C, Mühleisen H, Mueller W, Pohl U, Prinz M, Benner A, Zapatka M, Gottardo NG, Driever PH, Kramm CM, Müller HL, Rutkowski S, von Hoff K, Frühwald MC, Gnekow A, Fleischhack G, Tippelt S, Calaminus G, Monoranu C-M, Perry A, Jones C, Jacques TS, Radlwimmer B, Gessi M, Pietsch T, Schramm J, Schackert G, Westphal M, Reifenberger G, Wesseling P, Weller M, Collins VP, Blümcke I, Bendszus M, Debus J, Huang A, Jabado N, Northcott PA, Paulus W, Gajjar A, Robinson GW, Taylor MD, Jaunmuktane Z, Ryzhova M, Platten M, Unterberg A, Wick W, Karajannis MA, Mittelbronn M, Acker T, Hartmann C, Aldape K, Schüller U, Buslei R, Lichter P, Kool M, Herold-Mende C, Ellison DW, Hasselblatt M, Snuderl M, Brandner S, Korshunov A, von Deimling A, Pfister SM (2018) DNA methylation-based classification of central nervous system tumours. Nature 555(7697):469–474

21. Boulay G, Awad ME, Riggi N, Archer TC, Iyer S, Boonseng WE, Rossetti NE, Naigles B, Rengarajan S, Volorio A, Kim JC, Mesirov JP, Tamayo P, Pomeroy SL, Aryee MJ, Rivera MN (2017) OTX2 activity at distal regulatory elements shapes the chromatin landscape of group 3 medulloblastoma. Cancer Discov 7(3):288–301. https://doi.org/10.1158/2159-8290.CD-16-0844

22. Lee Y, Miller HL, Jensen P, Hernan R, Connelly M, Wetmore C, Zindy F, Roussel MF, Curran T, Gilbertson RJ, McKinnon PJ (2003) A molecular fingerprint for medulloblastoma. Cancer Res 63(17):5428–5437

23. Reynolds BA, Weiss S (1992) Generation of neurons and astrocytes from isolated cells of the adult mammalian central nervous system. Science 255(5052):1707–1710

24. Reynolds BA, Rietze RL (2005) Neural stem cells and neurospheres—re-evaluating the relationship. Nat Methods 2(5):333–336. https://doi.org/10.1038/nmeth758

25. Galli R (2013) The neurosphere assay applied to neural stem cells and cancer stem cells. Methods Mol Biol 986:267–277. https://doi.org/10.1007/978-1-62703-311-4_17

Chapter 4

Bioinformatic Strategies for the Genomic and Epigenomic Characterization of Brain Tumors

Vijay Ramaswamy and Michael D. Taylor

Abstract

Genomics has significantly advanced our knowledge of the biology of brain tumors, and refined our classification over the past 10 years. These advances have relied on the unbiased analysis of large cohorts of brain tumors, where they are clustered in an unbiased manner prior to ascribing clinical and biological features. Indeed, this has resulted in the identification of several layers of heterogeneity not previously appreciated by morphology alone. As such, the classification of unknown samples into known biological subgroups can be performed robustly using either gene expression or DNA methylation profiling.

Key words DNA methylation profiling, CpG sites, Hybridization, FFPE tissue, RNA sequencing, Integrated analysis, Copy number analysis, Pathway analysis

1 Introduction

The advent of integrated genomics has significantly enhanced our knowledge of the biology of brain tumors [1–3]. Transcriptional profiling was the first modality applied to the unbiased analysis of brain tumors, specifically medulloblastoma (MB) and glioblastoma (GBM). Indeed, the initial clustering of both GBM and MB in the early 2000s using gene expression arrays revealed that both entities were comprised of several distinct molecular variants despite identical morphological appearances [4–6]. Eventual integration of somatic copy number variants and single nucleotide variants revealed that MB comprises four distinct molecular subgroups termed WNT, SHH, Group 3, and Group 4 with distinct copy number profiles, mutational profiles, demographics, and outcomes [7–17]. The application of gene expression profiling to GBM provided similar insights, whereby integration of somatic nucleotide variants revealed that IDH1/2 mutated and 1p19q co-deleted glioma's represented distinct diseases, with specific responses to treatment [18–21]. Subsequent application of unbiased transcriptional profiling to other brain tumor entities including atypical

Sheila K. Singh and Chitra Venugopal (eds.), *Brain Tumor Stem Cells: Methods and Protocols*, Methods in Molecular Biology, vol. 1869, https://doi.org/10.1007/978-1-4939-8805-1_4, © Springer Science+Business Media, LLC, part of Springer Nature 2019

teratoid rhabdoid tumor (ATRT), CNS primitive neuroectodermal tumor (CNS-PNET), choroid plexus tumors, and ependymoma has revealed considerable heterogeneity across these entities as well [22–30]. The application of next generation sequencing, specifically DNA sequencing has allowed the characterization of the full spectrum of single nucleotide variants and structural rearrangements, and their subgroup specificity. Indeed, several subgroups of CNS-PNET and MB are characterized by recurrent fusions, duplications, and/or single nucleotide variants, such as mutations in the SHH pathway within MB subgroups. RNA sequencing is a more powerful modality for gene expression profiling, which allows for identification of fusion transcripts and alternate transcripts; however it requires much more computational expertise.

A limitation of gene expression profiling is the requirement of high quality RNA extracted from flash-frozen tissue. Recently, DNA methylation profiling has emerged as a robust alternative to gene expression-based classification mainly due to the increased availability of DNA from even formalin fixed paraffin embedded tissue. Indeed, in our experience, robust data can be generated from FFPE material as old as 50 years, and does not exhibit any significant batch effects. Another major advantage of DNA methylation profiling is the ability to infer relatively robust copy number profiles using the sum of the methylated and unmethylated probes. Broad arm-level copy number profiles can be robustly generated using DNA methylation arrays, and some highly covered focal amplifications such as MYC, MYCN, PDGFR, and CDK6 can be identified. However, several disadvantages exist with DNA methylation, specifically it tends to reflect cell of origin rather than dynamic cellular changes, and due to the nature of the array pathway analysis is biased and unreliable. However despite these limitations, DNA methylation, particularly the application of Illumina HumanMethylationEPIC arrays, has emerged as the leading platform for classification of brain tumors [29].

Recently, integrated methods have been applied to the classification of brain tumors, whereby DNA methylation is combined with gene expression[27, 28]. Many advantages exist with this method, particularly it harnesses the full spectrum of both data types, and recently we have shown that integrated DNA methylation with gene expression significantly improves classification compared to each modality individually. In this chapter, we will introduce protocols for processing of tissue for both gene expression and DNA methylation analysis, and we provide initial methods of preprocessing of data and consensus clustering methods.

2 Materials

2.1 Extraction of DNA and RNA

1. Trizol or equivalent TriReagent solution using the guanidinium thiocyanate-phenol-chloroform extraction (*see* **Note 1**).

2. Proteinase K (Stock Concentration 20 mg/mL).

3. DNA extraction buffer (Final concentrations of 10 mM NaCl, 20 mM Tris–HCl pH 8.0, 1 mM EDTA).

4. Phenol:chloroform (*see* **Note 2**).

5. Qiagen DNeasy FFPE extraction kit (Qiagen Cat No. 56,404) (*see* **Note 2**).

6. Tissue homogenizer (Precellys 24 Homogenizer or equivalent).

7. 1.5–2 mL microfuge tube compatible centrifuge, capable of speeds up to $15,000 \times g$.

8. Thermomixer.

9. RNAase AWAY wipes or spray.

10. High quality RNAase-free water.

11. EZ DNA Methylation™ Kit (Zymo Corporation, Irvine, CA).

2.2 DNA Methylation and Gene Expression Data

1. The raw methylation and expression data across 763 medulloblastoma samples (GEO GSE85218)(*see* **Note 4**).

2. A network server and if necessary a desktop computer (*see* **Note 3**).

3. Affymetrix or Illumina gene expression arrays (*see* **Note 5**) (RNA-sequencing can also be used depending on level of user (*see* **Note 6**)).

4. Platforms for DNA methylation profiling include Illumina HumanMethylationEPIC arrays, which allow the profiling of over 850,000 CpG sites and it is currently the only available array-based platform for DNA methylation profiling (*see* **Note 6**).

2.3 Preprocessing and Clustering Software

1. R statistical environment (https://www.r-project.org/) (*see* **Note 7**).

    (a) *Bioconductor* package (https://www.bioconductor.org).

    (b) *SNF* package (https://cran.r-project.org/web/packages/SNFtool/index.html).

    (c) *ConsensusClusterPlus* (http://bioconductor.org/packages/release/bioc/html/ConsensusClusterPlus.html).

2. Tuxedo Suite (Cufflinks, Bowtie, Tophat) to be run in Terminal (https://support.illumina.com/help/BS_App_RNASeq_Alignment_OLH_1000000006112/Content/Source/Informatics/Apps/TuxedoSuite_RNASeqTools.htm) or can be run through GenePattern (https://software.broadinstitute.org/cancer/software/genepattern/

3. Affymetrix Expression Console (ThermoFisher, Waltham, MA).

4. Partek Genomic Suite (Partek Corporation, St. Louis, MO).

5. Illumina Genome Studio (Illumina Inc., San Diego, CA).

2.4 Pathway Analysis Software

1. Gene Set Enrichment Analysis (GSEA, http://software.bro adinstitute.org/gsea/index.jsp) through the Broad Institute (*see* **Note 8**).

2. MSigDB is also a Broad Institute online tool (http://software. broadinstitute.org/gsea/msigdb) (*see* **Note 9**).

3. g:profiler website (biit.cs.ut.ee/gprofiler/) (*see* **Note 10**).

3 Methods

There are three parts to the protocol. First is appropriate processing of nucleic acids from either frozen or FFPE-derived tissue for hybridization to either gene expression arrays or DNA methylation. Second is the preprocessing of gene expression or DNA methylation data for downstream applications. Finally, we will provide basic strategies for exploratory data analysis using either unbiased clustering or pathway analysis.

3.1 Nucleic Acid Extraction

3.1.1 RNA Extraction from Frozen Tissue

Ideally RNA extraction should be performed using the TriZol method, as all RNA moieties including microRNA and other noncoding RNA will be isolated for downstream analysis (*see* **Note 11**). Proper extraction with TriZol will typically yield very high quality RNA. Column-based extraction methods can be used to isolate high quality RNA; however in general small moieties under 100 bp will not be reliably isolated (*see* **Note 11**). RNA extraction from cell culture can be performed with TriZol added directly to a tissue culture plate or centrifuged cells at **step 3**.

1. A pea sized portion of tissue should be snap frozen in liquid nitrogen upon collection. The sample can be cut on dry ice if DNA extraction is also required.

2. Cool a mortar and pestle with liquid nitrogen. Add tissue sample along with a few milliliters of liquid nitrogen, and grind tissue as finely as possible. Transfer to a 2 mL centrifuge tube. A tissue homogenizer can also be used when frozen tissue quantities are small after suspension of the frozen tissue in Trizol. Ensure use of screw top caps to avoid loss of Trizol.

3. Add 1 mL of Trizol (ThermoFisher, or similar Tri-Reagent solution). Vortex until most of the tissue is broken up.

4. Incubate for 5 min on ice.

5. Add 200 µL of chloroform and shake for 15 s (do not vortex), and allow the mixture to sit for 2–5 min on ice.

6. Centrifuge samples at an rcf of 12,000 × g for 15 min at 4 °C.

7. Prepare new sterile 2 mL centrifuge tubes with 500 µL of 2-propanol, and add supernatant (top layer above white precipitate containing RNA) from the centrifugation step to the labeled tubes of 2-propanol. Mix thoroughly by gently flicking the tubes and allow RNA to precipitate on ice for 10 min. If small amounts of tissue are used, glycogen can be added as an RNA carrier.

8. Centrifuge at 12,000 rpm for 15 min.

9. Discard the supernatant and gently wash pellet with 1 mL of 70% ethanol (*see* **Note 12**).

10. Centrifuge at an rcf of >12,000 × g for 5 min.

11. Discard the supernatant, invert the tube, and allow to drain on a paper towel for 5–10 min.

12. Resuspend pellet in 20–40 µL of RNAase-free water and freeze at −80 °C (*see* **Note 13**).

13. Assess RNA quality using a bioanalyzer. RNA integrity above seven is required for array-based hybridization. If assessing by gel electrophoresis, 500 ng of product should be run on a 1% agarose gel, and three clear bands will be observed representing the ribosomal RNA subunits (28S, 18S, and 5S (*see* **Note 14**)).

3.1.2 DNA Isolation from Frozen Tissue

A pea sized portion of frozen tissue should be used for this process. Tissues can either be incubated with lysis buffer and proteinase K overnight or alternatively frozen tissue from the RNA extraction can be finely ground using a mortar and pestle and incubated for 3 h. It is desirable to process small tissue quantities using the overnight digestion method without grinding (*see* **Note 3**). DNA isolation from cells grown in culture involves the same steps with the exception of adding lysis buffer and proteinase K directly to centrifuged cells.

1. Add 920 µL of DNA extraction buffer (10 mM NaCl, 20 mM Tris–HCl pH 8.0, 1 mM EDTA), 50 µL of 10% SDS (final concentration of 0.5%), and 30 µL of 10 mg/mL proteinase K solution to the frozen tissue and incubate at 55 °C on a thermomixer from between 3 h to overnight. If tissue is not first minced, overnight digestion is preferred. The lysed tissue will be relatively clear.

2. Optional: 10–15 microliters of RNAase (20 mg/mL stock) can be added at this step and incubated at 37 °C for 15 min.

3. Add an equal volume of equilibrated phenol, close cap tightly and mix gently by inversion. Do NOT vortex to avoid shearing of genomic DNA.

4. Centrifuge the tube at 2000 × *g* in a swinging bucket rotor at room temperature for 10 min. Using a plastic pipette, carefully transfer the upper clear aqueous layer to a clean tube.

5. Repeat phenol extraction from **steps 3** and **4** one more time.

6. To remove the phenol, extract with an equal volume of chloroform, centrifuge for 10 min at room temperature at 2000 × *g* and carefully transfer the supernatant to a clean tube. This can be repeated twice to eliminate all phenol from the DNA sample.

7. To the aqueous DNA sample, add 0.1 vol of 3 M sodium acetate (pH 5.2) and mix by flicking tube. Add 2 volumes of ice cold 100% ethanol to the tube and mix by inversion. A cotton-like precipitate should form. Using a sterile plastic rod, spool the precipitated DNA and rinse the spooled DNA with 1 mL of cold 70% ethanol. If no precipitate is observed, centrifuge the sample at full speed for 10 min and wash pellet with 70% ethanol.

8. Air dry for 10–15 min and resuspend in 100–500 μL of TE buffer (10 mM Tris–HCl, pH 8.0 and 0.1 mM EDTA) (*see* **Note 12**).

9. Analyze integrity by gel electrophoresis where a large band at 12,000 kb should be visible or analyze DNA integrity using a bioanalyzer.

3.1.3 DNA Isolation from FFPE Tissue

DNA isolation from FFPE tissue should be performed using the Qiagen FFPE DNeasy kit, or the FFPE All-Prep Kit which allows for simultaneous extraction of RNA. 5–10 unstained slides or the equivalent in scrolls should be used. If ample tissue quantities are present on the slide, then 3–5 slides may be used.

1. To remove tissue from the slides, use a sharp sterile razor blade, and at a 45° angle scrape the tissue off the slide in one motion. The same blade should be used for multiple slides as the tissue will begin to curl allowing easy transfer to a 2 mL centrifuge tube.

2. Add 1 mL of xylene to the tissue to deparaffinize, allow it to sit at room temperature for 5 min and then centrifuge at an rcf of 12,000 × *g* for 5 min. Remove xylene using a plastic 1 mL pipette tip and repeat xylene incubation. If the tissue is not strictly adherent to the microfuge tube, xylene can be left in the tube, as the next ethanol step will remove it.

3. Add 1 mL of 100% ethanol to dry out xylene and vortex briefly. Incubate for 5 min and then centrifuge at an rcf of 12,000 × *g* for 5 min. Carefully remove all traces of ethanol using a long pipette tip being careful to not disturb the tissue. Repeat once.

4. Dry samples for 10 min at 37 °C.

5. Proceed with DNA extraction as per Qiagen protocol.

6. At the end of the extraction, resuspend DNA in 100 μL of TE buffer and store at 4 °C for immediate use or −20 °C for long-term storage.

3.2 Gene Expression Profiling

Sample preparation and preprocessing will be described for Affymetrix chips; however, a similar method is available should Illumina beadchip arrays be used. Sample preparation is almost always performed in core facilities as per the manufacturer's instructions. Gene expression profiling is particularly vulnerable to batch effects, and samples should be ideally submitted in one batch, in a random order. If multiple batches cannot be avoided, strategies exist to correct batch effects such as the Combat package. If Illumina beadchip arrays are used, allocation of samples to the arrays should be random as each array has 12 positions. The goal of preprocessing of gene expression data is to first normalize the data, and convert probe intensities to relative expression values. Open source methods to perform preprocessing and normalization include the arrays workflow within the R statistical environment (https://www.bio conductor.org/help/workflows/arrays/). The arrays workflow leverages bioconductor, and the affy package for array preprocessing and the limma package for two-color preprocessing and differential expression. To process affymetrix arrays, two files are required. CEL files contain measured intensities and locations for an array that has been hybridized. There is one .cel file per sample. A CDF file is provided by Affymetrix and provides the layout of the chip.

1. 50–500 ng of RNA with an RNA integrity number (RIN) above seven are required for both Affymetrix and Illumina gene expression arrays. Submit 500 ng of RNA ideally at a concentration of at least 40 ng/μL (*see* **Notes 15** and **16**).

2. Sample preparation and hybridization is performed by core facilities as per the manufacturer's instructions. Quality control of individual samples is also typically done within core facilities, and as such they should be consulted prior to initiation of the experiment.

3. Preprocessing and normalization can be performed in two methods. Commercial software from Affymetrix can be used to generate normalized expression and instructions are as follows https://tools.thermofisher.com/content/sfs/manuals/expression_console_userguide.pdf. Bioconductor can be used for preprocessing and generation of differential expression using the arrays package which includes the affy and limma packages. The lumi package along with the illumina annotations files can be used to process illumine-based arrays.

4. Two methods of preprocessing are widely used. Preprocessing converts intensities to expression values, and is also called normalization. MAS5 is an Affymetrix proprietary method which provides usable data with single chips, and can be nearly reproduced using the affy package in bioconductor. The MAS5 method relies heavily on mismatch probes on the chip. RMA (Robust multichip analysis) assumes all chips have some background distribution of values, and does not use the mismatch probes which decreases accuracy but increases precision. RMA does not provide p-values for each probe as MAS5 does; however, it provides a robust model, where the model is defined and then it fits the experimental data to the model. RMA provides log2 transformed values, and requires multiple samples.

5. Either the *affy* and *limma* packages can be used to generate normalized expression data, or alternatively expression console can be used; however for downstream analysis data should be log2 transformed to ensure that variance is similar across orders of magnitude (a twofold decrease is the same as a twofold increase).

6. If RNA-sequencing is used, several packages exist to align, normalize, and generate transcript counts; however, the most widely used is the Tuxedo suite, where Tophat is used to align and map the transcripts and Cufflinks used to generate differentially expressed genes. RNA-sequencing can be aligned and preprocessed using either a high performance server or using the GenePattern or Galaxy (https://usegalaxy.org/) online suites.

7. Once data is normalized, generate list of differentially expressed genes by median absolute deviation. Usually the top 1–2% differentially expressed genes are used for downstream analysis. The top differentially expressed genes are used for analysis to eliminate uninformative probes which do not vary across the experiment. This can be performed within the R-statistical environment using the following command or within Microsoft Excel using the stdev and sort functions:

Assuming the data has been loaded into an object termed "testset," then the following will sort the data based on decreasing variance and only select the top 200 differentially expressed genes:

```
mads=apply(testset,1,mad)
beta.values<-beta.values[rev(order(mads))[1:300],]
```

For downstream consensus clustering data should be ideally median centered particularly for k-means consensus clustering.

3.3 DNA Methylation Profiling

The processing of DNA methylation arrays requires bisulfite conversion of DNA, which is a process whereby unmethylated cytosines are deaminated to produce uracil in DNA. Methylated cytosines are protected from the conversion to uracil, allowing the use of direct sequencing to determine the locations of unmethylated cytosines and 5-methylcytosines at single nucleotide resolution or hybridization to Illumina Methylation arrays which have specific probes to both methylated and unmethylated CpG sites. For the purposes of this chapter only array-based DNA methylation analysis will be discussed.

1. At least 200 ng of Picogreen quantified DNA (either FFPE derived or Frozen) should be bisulfite converted using the Zymo EZ Methylation kit as per the manufacturer's instructions. The conversion process should utilize the protocol whereby 16 cycles of 30 s at 95 °C followed by 60 min at 50 °C followed by a hold at 4 °C should be used. The remainder of the protocol is as per the manufacturer's instructions (http://www.zymoresearch.com/downloads/dl/file/id/56/d5001i.pdf). Less than 200 ng can be used but it is not recommended by the manufacturer (*see* **Note 17**).

2. For FFPE-derived samples the Illumina FFPE QC assay should be run first prior to bisulfite conversion, and only samples with less than a 5 cycle difference from the supplied control should be used for analysis.

3. The FFPE restore kit is required for FFPE-derived DNA, and is typically performed at the core facility along with the hybridization.

4. Once the samples have been hybridized, two files are available, a green channel and a red channel. These represent unmethylated and methylated probes respectively.

5. Samples can be normalized either using Illumina Genome Studio as per the manufacturer's instructions, or using the open source R statistical environment. A one command analysis pipeline to load, normalize, and generate differentially methylated probes is the ChAMP package (https://bioconductor.org/packages/release/bioc/html/ChAMP.html). Beta values are calculated as the fraction of the probe that is methylated (methylated/(methylated+unmethylated)). A matrix of beta values is analogous to a matrix of relative gene expression values.

6. Differentially methylated probes (usually the top 5000–10,000 probes) are used for downstream analysis and can be calculated using the probes with the top median absolute deviation.

3.4 Initial Exploration of Data

Initial exploration of data depends on the type of experiment. If a large number of tumors have been profiled either by gene expression or DNA methylation, then the top differentially expressed or methylated probes are clustered, using a variety of methods. The more the samples available, the more robust the exploration of data will be;

however it will also be more time consuming. Consensus clustering, and negative matrix factorization can both be performed either using the open source R packages *consensusclusterplus* or *NMF* respectively, or within the online tool GenePattern from the Broad Institute (http://software.broadinstitute.org/cancer/software/genepattern/). It should be noted that NMF is extremely resource intensive, and requires several days to perform when sampling over 100 samples (*see* **Note 18**)). The Gene Pattern users guide provides detailed instructions on how to format data for analysis.

1. Samples can be clustered using either the consensusclusterplus package or by using the hierarchical clustering which is built into the bioconductor package. Kmeans and Hierarchical clustering are two commonly applied clustering methods, whereby samples are clustered into 2–20 groups. For example, using the medulloblastoma dataset (GSE GSE85218), $K = 4$ will yield groups with demographics and copy number profiles similar to the four medulloblastoma subgroups using either DNA methylation or gene expression. For gene expression typically the top 50–300 differentially expressed genes are used, and for DNA methylation the top 1000–10,000 differentially methylated probes are used. The analysis should be repeated using varying numbers of probes.

2. Unsupervised hierarchical clustering can be used as an orthogonal method to consensus clustering. Programs such as pvclust can help ascribe a p-value to separations in the dendrogram to help determine the true difference between samples.

3. Negative matrix factorization is a third orthogonal method for exploration of biologically discrete groups.

4. Principal component analysis and tSNE (*see* **Note 18**) (*t*-distributed stochastic neighbor embedding, a machine learning algorithm for dimensionality reduction) can also be employed as effective visual tools for exploratory data analysis and to help determine the degree of heterogeneity across samples.

5. The optimal number of relevant clusters/groups can be determined in an unbiased manner exploiting several unsupervised methods. If using consensus clustering, then the change in the Gini coefficient can be used to help determine the optimal number of statistically discrete groups. If using NMF, a cophenetic coefficient will be generated whereby the biggest drop usually preceded by an increase in cophenetic coefficient represents the most robust number of groups. The groups determined using each modality independently should be compared, and usually relevant groups will be similar regardless of the method chosen. Newer density based methods such as DBSCAN (Density-based spatial clustering of applications with noise) (r package dbscan) or spectral clustering allows an orthogonal method of identifying relevant clusters in an unbiased manner [30].

6. An additional method of classifying samples once consensus clustering has been performed is silhouette width analysis, which is a succinct graphical representation of how well an object lies within its cluster. Samples with negative silhouette widths are typically discarded from the analysis.

3.5 Assignment of Subgroups

Subgroups for known entities such as medulloblastoma, glioblastoma, ependymoma, and ATRT can be assigned to an unknown dataset using machine learning algorithms. The most popular and easiest to use is PAM (Prediction Analysis of Microarrays). Either an excel plugin (http://statweb.stanford.edu/~tibs/PAM/) can be used, or alternatively the pamr package in the R statistical environment can be used.

1. Using previously generated normalized data, two matrices are created, one is the training set and one is the test set. The training set consists of known samples which have been previously subgrouped. Using the top 25–500 differentially expressed transcripts, or the top 1000–1000 differentially methylated probes, a tab delineated file is created whereby the first column is the gene name, and each subsequent column is the expression or betavalues for that particular gene. Identical normalization procedures should be used for both training and test sets, and ideally the same platform should be employed. The training set has an additional row underneath the sample identifier which contains the group previously assigned to the sample.

2. For several entities, training sets are freely available to download, including a 25 gene classifier that robustly classifies medulloblastoma samples into the four subgroups. For entities such as ependymoma, ATRT, and glioblastoma, several datasets are available to train the classifier.

3. If the pamr package is being used, the following simple script can be employed if a robust training set is available.

   #import the training set and train pamr. The 103 refers to the number of samples in the training set.
   train.data   <-pamr.from.excel("train.txt",   103,   sample.labels=TRUE)
   train.train <-pamr.train(train.data)

   #import the test set. The 14 refers to the number of samples in the test dataset
   test.data   <-pamr.from.excel("test.txt",   14,   sample.labels=FALSE)

   test.predict<-pamr.predict(train.train,   test.data$x,   threshold=1, type="posterior")
   test.class<-pamr.predict(train.train, test.data$x, threshold=1, type="class")

```
#write table
write.table(cbind(test.data$y, test.predict, test.class), "file-
name.txt", sep="\t", row.names=TRUE, col.names=NA);
```

4. A less robust method of class prediction is to cluster the unknown samples along with known samples. This can be done by normalizing the training and test datasets together, and then performing either consensus clustering or unsupervised clustering on the remaining matrix and determining which group the unknown samples cluster with. Batch effects need to be considered if using this method, and the training set should be sufficiently large to capture all relevant subgroups.

5. The Molecular Neuropathology 2.0 algorithm also known as the Heidelberg Brain Tumor Classifier can be used to predict known brain tumour entities and subgroups using DNA methylation profiling using the online tool (https://www. molecularneuropathology.org/mnp). Both 450 and 850 k methylation arrays can be used with this package and this tool has been validated in a robust manner across >4000 samples [29] (*see* **Note 19**).

3.6 Integrated Analysis

In Subheading 3.4, methods for the analysis of individual data types are presented. Several methods exist to aggregate various data types. Aggregating multiple data types confers several advantages, whereby unique information across each data type can provide unique contributions. An emerging powerful method to combine multiple data types is similarity network fusion (SNF). This method does not rely on preselection of genes, and instead creates a fused patient network whereby the entire spectrum of data is used. This has the distinct advantage of providing shared and complementary information from each data type. The advantage of integrating data types using SNF is that weak similarities (low-weight edges) disappear, which helps to reduce the noise, and strong similarities (high weight edges) present in multiple networks networks are added to the others. In addition, low-weight edges supported by all networks are retained depending on how tightly connected they are across networks [27]. We have recently applied SNF followed by spectral clustering to robustly identify the full spectrum of heterogeneity across medulloblastoma subgroups. Indeed, we showed that combining gene expression and DNA methylation across 763 primary medulloblastoma samples, is able to delineate substructure within medulloblastoma subgroups more robustly than each data type on its own. The full methods of applying SNF with spectral clustering have been described by Wang et al., and Cavalli et al.; and can be applied using the SNF package in the R statistical environment (https://cran.r-project.org/web/packages/SNFtool/index.html) [27, 28].

3.7 Copy Number Analysis

1. Copy number analysis can be performed using dedicated copy number arrays from Affymetrix and Illumina (Affymetrix SNP 6.0/Cytoscan or Illumina Omni chips). Illumina and Affymetrix provide software to quickly process files. Diploid controls are required and available from several published datasets.

2. DNA copy number can also be inferred from DNA methylation arrays. The *conumee* package is a very useful tool to infer copy number from Illumina DNA methylation arrays. The basic premise of this method is to calculate the sum of the intensities of the methylated and unmethylated probes, and treat the resulting combined intensity value as a copy number array. Normal tissue controls are available through the Cancer Genome Atlas if the Infinium HumanMethyation450 arrays are being used; however if the Infinium HumanMethylationE-PIC platform is being used, normal controls will ideally need to be run. Ideally 20–30 diploid controls should be used for this analysis. The conumee package provides detailed instructions on generation of copy number profiles, and has robust performance from methylation arrays generated from both frozen and FFPE-derived tissue (*see* **Notes 20** and **21**).

3. Whole genome sequencing can be used to detect very specific breakpoints but is beyond the scope of this chapter.

3.8 Pathway Analysis

Multiple tools for pathway analysis can be used. First, using a combination of consensus clustering, SNF and/or copy number profiles, the optimal number of discrete groups is determined, and by calculating the top differentially expressed genes, the signature genes for each group discerned can be determined. GSEA can be either downloaded onto a desktop computer, or a local server can be used. G:profiler and MSigdb are both online tools.

1. If using Geneset Enrichment Analysis (GSEA), then no pre-ranking or preselection of genes is required, and a matrix of the entire data set can be converted to a .gct file, while a .cls file is generated that assigns each sample to a group to be compared. Instructions on how to create .gct and .cls files are present on the GSEA wiki page (http://software.bro adinstitute.org/cancer/software/gsea/wiki/index.php/ Data_formats).

2. Once the .gct and .cls files are generated, these can be loaded into the GSEA software for analysis.

3. The C5 and C6 genesets are typically what are used for the analysis of brain tumors (C5 genesets are GO terms and C6 are previously generated microarray datasets).

4. g:profiler can also be used for pathway analysis, and provides robust visualization. The dataset should be pre-ranked with differentially expressed genes used for the analysis. Usually a

cutoff of a corrected *p*-value of 0.05 with a log2 fold change of 2 is used to select genes specific to a particular dataset. Alternatively, features genes generated by SNF and spectral clustering can be used to identify signature genes through integrative clustering. Pre-ranked genes in decreasing order of significance are then pasted into the g:profiler online portal (http://biit.cs.ut.ee/gprofiler/).

5. MSigDb is similar to GSEA, whereby curated datasets by the broad institute are used, however a pre-ranked geneset list can be used.

4 Notes

1. Phenol:chloroform should be fresh and used within 2 months of opening.

2. Qiagen All-Prep kits can also be used to obtain RNA, DNA, and microRNA from frozen or FFPE in single reaction. Yield is identical to using kits individually. AllPrep DNA/RNA/miRNA Universal Kit (Cat 80,224) can be used in lieu of Trizol and Proteinase K to extract DNA, RNA and miRNA from a single reaction.

3. A network server is required for this analysis, however once the data is preprocessed, a list of differentially expressed genes and differentially methylated probes may be analyzed on a desktop computer.

4. User generated gene expression data can be obtained from either Affymetrix or Illumina platforms, and both Affymetrix Expression Console and Illumina GenomeStudio software allows generation of differentially expressed genes and differentially methylated probes for downstream application. RNA sequencing is emerging as a more robust method allowing for digital transcript counts, fusion detection, and SNV calling. RNA sequencing also has less batch effects. If cost and expertise allow, RNA sequencing can generate more robust gene expression data for downstream analysis.

5. In the selection of expression array platforms, the Cancer Genome Atlas has used Affymetrix U133p2 arrays, and this may represent a reasonable choice to allow normalization with existing data. Moreover, Affymetrix U133p2 arrays can be normalized using the online R2 tool (https://hgserver1.amc.nl/cgi-bin/r2/main.cgi) in a very simple and straightforward manner.

6. Digital next generation sequencing methods including methyl capture sequencing, and whole genome bisulfite sequencing are emerging, however are in their infancy and not widely applied.

7. Both open source (R statistical environment) and commercial methods (Partek Genomic Suite, Partek Corporation, St. Louis, MO) provide basic bioinformatics analysis options, whereby lists of differentially expressed genes can be used for downstream analysis.

8. GSEA provides a very easy-to-use interface, and does not require pre-ranking of genes, as the software performs its own ranking of genes. The GSEA database is consistently curated and updated.

9. MSigDB is also a Broad Institute online tool using the same curated genesets as GSEA, which allows a ranked list of genes to be inputted and pathway determined.

10. G:profiler is an online tool whereby ordered lists of decreasing importance can be analyzed using an easy-to-use web-based format. It is a public web server for characterizing and manipulating gene lists of high-throughput genomics. G:Profiler has a simple web interface with powerful visualization, with the advantage that it is typically updated every 2 months in sync with Ensembl.

11. Some column-based kits allow for isolation of microRNA's under 30 bp, which can be used for downstream analysis.

12. RNA extraction should always proceed under RNAase-free conditions, including use of RNase AWAY spray and wipes (Available from most major suppliers including VWR, ThermoFisher).

13. The pellet may not be visible and microfuge tubes should be centrifuged in the same orientation to ensure RNA or DNA are on the same side of the tube.

14. Intense bands observed in the wells, or a 12 kb band indicates presence of DNA contamination. Similarly, additional peaks beyond the three ribosomal RNA subunits on a bioanalyzer may represent DNA contamination or RNA degradation.

15. Lower amounts of RNA can be used, but should be discussed with the core facility prior to proceeding.

16. Choice of expression profiling platform should depend on existing data available for the experiment for direct comparison. In general, for human gene expression, Affymetrix U133p2 allows direct comparison to hundreds of other datasets including normal tissue as part of the Cancer Genome Atlas.

17. As little as 25 ng of DNA can be used for bisulfite conversion and hybridization, but relies heavily on the quality of the DNA and the experiment to be performed. If simple clustering, then small amounts can be used. Successful hybridization cannot be

guaranteed at such low concentrations, and if possible it should be avoided.

18. The NMF package in the R Statistical Environment can also be used when run on a Network Server, and is generally more reliable for larger datasets. Running this package on a desktop or laptop computer can lead to severe damage, and should not be performed.

19. tSNE visualization can be performed using the Rtsne package in the R statistical environment, and ideally should be the sum of at least 10,000 iterations. Perplexity values between 5 and 50 should be explored.

20. The MNP online tool allows both prediction of the entity (e.g., medulloblastoma, glioblastoma, ependymoma, ATRT) as well as individual subgroups such as the four medulloblastoma subgroups. Both FFPE and frozen-derived samples can be used, and copy number profiles will be generated using the sum of the methylated and unmethylated probes.

21. http://bioconductor.org/packages/release/bioc/html/con umee.html—The conumee package should be run as per the vignette using default settings. Copy number can be inferred by direct visualization of the genome.

References

1. Ramaswamy V, Taylor MD (2017) Medullo-blastoma: from myth to molecular. J Clin Oncol 35(21):2355–2363. https://doi.org/10.1200/jco.2017.72.7842

2. Jones C, Karajannis MA, Jones DT, Kieran MW, Monje M, Baker SJ, Becher OJ, Cho YJ, Gupta N, Hawkins C, Hargrave D, Haas-Kogan DA, Jabado N, Li XN, Mueller S, Nicolaides T, Packer RJ, Persson AI, Phillips JJ, Simonds EF, Stafford JM, Tang Y, Pfister SM, Weiss WA (2016) Pediatric high-grade glioma: biologically and clinically in need of new thinking. Neuro-Oncology. https://doi.org/10.1093/neuonc/now101

3. Louis DN, Perry A, Reifenberger G, von Deimling A, Figarella-Branger D, Cavenee WK, Ohgaki H, Wiestler OD, Kleihues P, Ellison DW (2016) The 2016 World Health Organization classification of tumors of the central nervous system: a summary. Acta Neuropathol 131(6):803–820. https://doi.org/10.1007/s00401-016-1545-1

4. Pomeroy SL, Tamayo P, Gaasenbeek M, Sturla LM, Angelo M, McLaughlin ME, Kim JYH, Goumnerova LC, Black PM, Lau C, Allen JC, Zagzag D, Olson JM, Curran T, Wetmore C, Biegel JA, Poggio T, Mukherjee S, Rifkin R, Califano A, Stolovitzky G, Louis DN, Mesirov JP, Lander ES, Golub TR (2002) Prediction of central nervous system embryonal tumour outcome based on gene expression. Nature 415(6870):436–442. https://doi.org/10.1038/415436a

5. Phillips HS, Kharbanda S, Chen R, Forrest WF, Soriano RH, Wu TD, Misra A, Nigro JM, Colman H, Soroceanu L, Williams PM, Modrusan Z, Feuerstein BG, Aldape K (2006) Molecular subclasses of high-grade glioma predict prognosis, delineate a pattern of disease progression, and resemble stages in neurogenesis. Cancer Cell 9(3):157–173. https://doi.org/10.1016/j.ccr.2006.02.019

6. Thompson MC, Fuller C, Hogg TL, Dalton J, Finkelstein D, Lau CC, Chintagumpala M, Adesina A, Ashley DM, Kellie SJ, Taylor MD, Curran T, Gajjar A, Gilbertson RJ (2006) Genomics identifies medulloblastoma subgroups that are enriched for specific genetic alterations. J Clin Oncol 24(12):1924–1931. https://doi.org/10.1200/JCO.2005.04.4974

7. Northcott PA, Korshunov A, Witt H, Hielscher T, Eberhart CG, Mack S, Bouffet E, Clifford SC, Hawkins CE, French P, Rutka JT,

Pfister S, Taylor MD (2011) Medulloblastoma comprises four distinct molecular variants. J Clin Oncol 29(11):1408–1414. https://doi.org/10.1200/JCO.2009.27.4324

8. Kool M, Koster J, Bunt J, Hasselt NE, Lakeman A, van Sluis P, Troost D, van Meeteren NS, Caron HN, Cloos J, Mrsić A, Ylstra B, Grajkowska W, Hartmann W, Pietsch T, Ellison D, Clifford SC, Versteeg R (2008) Integrated genomics identifies five medulloblastoma subtypes with distinct genetic profiles, pathway signatures and clinicopathological features. PLoS One 3(8): e3088. https://doi.org/10.1371/journal.pone.0003088

9. Taylor MD, Northcott PA, Korshunov A, Remke M, Cho YJ, Clifford SC, Eberhart CG, Parsons DW, Rutkowski S, Gajjar A, Ellison DW, Lichter P, Gilbertson RJ, Pomeroy SL, Kool M, Pfister SM (2012) Molecular subgroups of medulloblastoma: the current consensus. Acta Neuropathol 123(4):465–472. https://doi.org/10.1007/s00401-011-0922-z

10. Kool M, Korshunov A, Remke M, Jones DT, Schlanstein M, Northcott PA, Cho YJ, Koster J, Schouten-van Meeteren A, van Vuurden D, Clifford SC, Pietsch T, von Bueren AO, Rutkowski S, McCabe M, Collins VP, Backlund ML, Haberler C, Bourdeaut F, Delattre O, Doz F, Ellison DW, Gilbertson RJ, Pomeroy SL, Taylor MD, Lichter P, Pfister SM (2012) Molecular subgroups of medulloblastoma: an international meta-analysis of transcriptome, genetic aberrations, and clinical data of WNT, SHH, Group 3, and Group 4 medulloblastomas. Acta Neuropathol 123 (4):473–484. https://doi.org/10.1007/s00401-012-0958-8

11. Remke M, Hielscher T, Korshunov A, Northcott PA, Bender S, Kool M, Westermann F, Benner A, Cin H, Ryzhova M, Sturm D, Witt H, Haag D, Toedt G, Wittmann A, Schottler A, von Bueren AO, von Deimling A, Rutkowski S, Scheurlen W, Kulozik AE, Taylor MD, Lichter P, Pfister SM (2011) FSTL5 is a marker of poor prognosis in non-WNT/non-SHH medulloblastoma. J Clin Oncol 29 (29):3852–3861. https://doi.org/10.1200/jco.2011.36.2798

12. Cho Y-J, Tsherniak A, Tamayo P, Santagata S, Ligon A, Greulich H, Berhoukim R, Amani V, Goumnerova L, Eberhart CG, Lau CC, Olson JM, Gilbertson RJ, Gajjar A, Delattre O, Kool M, Ligon K, Meyerson M, Mesirov JP, Pomeroy SL (2011) Integrative genomic analysis of medulloblastoma identifies a molecular subgroup that drives poor clinical outcome. J Clin Oncol 29(11):1424–1430. https://doi.org/10.1200/JCO.2010.28.5148

13. Pugh TJ, Weeraratne SD, Archer TC, Pomeranz Krummel DA, Auclair D, Bochicchio J, Carneiro MO, Carter SL, Cibulskis K, Erlich RL, Greulich H, Lawrence MS, Lennon NJ, McKenna A, Meldrim J, Ramos AH, Ross MG, Russ C, Shefler E, Sivachenko A, Sogoloff B, Stojanov P, Tamayo P, Mesirov JP, Amani V, Teider N, Sengupta S, Francois JP, Northcott PA, Taylor MD, Yu F, Crabtree GR, Kautzman AG, Gabriel SB, Getz G, Jager N, Jones DT, Lichter P, Pfister SM, Roberts TM, Meyerson M, Pomeroy SL, Cho YJ (2012) Medulloblastoma exome sequencing uncovers subtype-specific somatic mutations. Nature 488(7409):106–110. https://doi.org/10.1038/nature11329

14. Jones DT, Jager N, Kool M, Zichner T, Hutter B, Sultan M, Cho YJ, Pugh TJ, Hovestadt V, Stutz AM, Rausch T, Warnatz HJ, Ryzhova M, Bender S, Sturm D, Pleier S, Cin H, Pfaff E, Sieber L, Wittmann A, Remke M, Witt H, Hutter S, Tzaridis T, Weischenfeldt J, Raeder B, Avci M, Amstislavskiy V, Zapatka M, Weber UD, Wang Q, Lasitschka B, Bartholomae CC, Schmidt M, von Kalle C, Ast V, Lawerenz C, Eils J, Kabbe R, Benes V, van Sluis P, Koster J, Volckmann R, Shih D, Betts MJ, Russell RB, Coco S, Tonini GP, Schuller U, Hans V, Graf N, Kim YJ, Monoranu C, Roggendorf W, Unterberg A, Herold-Mende C, Milde T, Kulozik AE, von Deimling A, Witt O, Maass E, Rossler J, Ebinger M, Schuhmann MU, Fruhwald MC, Hasselblatt M, Jabado N, Rutkowski S, von Bueren AO, Williamson D, Clifford SC, McCabe MG, Collins VP, Wolf S, Wiemann S, Lehrach H, Brors B, Scheurlen W, Felsberg J, Reifenberger G, Northcott PA, Taylor MD, Meyerson M, Pomeroy SL, Yaspo ML, Korbel JO, Korshunov A, Eils R, Pfister SM, Lichter P (2012) Dissecting the genomic complexity underlying medulloblastoma. Nature 488 (7409):100–105. https://doi.org/10.1038/nature11284

15. Thompson EM, Hielscher T, Bouffet E, Remke M, Luu B, Gururangan S, McLendon RE, Bigner DD, Lipp ES, Perreault S, Cho YJ, Grant G, Kim SK, Lee JY, Rao AA, Giannini C, Li KK, Ng HK, Yao Y, Kumabe T, Tominaga T, Grajkowska WA, Perek-Polnik M, Low DC, Seow WT, Chang KT, Mora J, Pollack IF, Hamilton RL, Leary S, Moore AS, Ingram WJ, Hallahan AR, Jouvet A, Fevre-Montange M, Vasiljevic A, Faure-Conter C, Shofuda T, Kagawa N, Hashimoto N, Jabado N, Weil AG, Gayden T, Wataya T, Shalaby T, Grotzer M,

Zitterbart K, Sterba J, Kren L, Hortobagyi T, Klekner A, Laszlo B, Pocza T, Hauser P, Schuller U, Jung S, Jang WY, French PJ, Kros JM, van Veelen MC, Massimi L, Leonard JR, Rubin JB, Vibhakar R, Chambless LB, Cooper MK, Thompson RC, Faria CC, Carvalho A, Nunes S, Pimentel J, Fan X, Muraszko KM, Lopez-Aguilar E, Lyden D, Garzia L, Shih DJ, Kijima N, Schneider C, Adamski J, Northcott PA, Kool M, Jones DT, Chan JA, Nikolic A, Garre ML, Van Meir EG, Osuka S, Olson JJ, Jahangiri A, Castro BA, Gupta N, Weiss WA, Moxon-Emre I, Mabbott DJ, Lassaletta A, Hawkins CE, Tabori U, Drake J, Kulkarni A, Dirks P, Rutka JT, Korshunov A, Pfister SM, Packer RJ, Ramaswamy V, Taylor MD (2016) Prognostic value of medulloblastoma extent of resection after accounting for molecular subgroup: a retrospective integrated clinical and molecular analysis. Lancet Oncol 7 (4):484–495. https://doi.org/10.1016/s1470-2045(15)00581-1

16. Ramaswamy V, Remke M, Adamski J, Bartels U, Tabori U, Wang X, Huang A, Hawkins C, Mabbott D, Laperriere N, Taylor MD, Bouffet E (2016) Medulloblastoma subgroup-specific outcomes in irradiated children: who are the true high-risk patients? Neuro-Oncology 18(2):291–297. https://doi.org/10.1093/neuonc/nou357

17. Lafay-Cousin L, Smith A, Chi SN, Wells E, Madden J, Margol A, Ramaswamy V, Finlay J, Taylor MD, Dhall G, Strother D, Kieran MW, Foreman NK, Packer RJ, Bouffet E (2016) Clinical, pathological, and molecular characterization of infant medulloblastomas treated with sequential high-dose chemotherapy. Pediatr Blood Cancer 63(9):1527–1534. https://doi.org/10.1002/pbc.26042

18. Eckel-Passow JE, Lachance DH, Molinaro AM, Walsh KM, Decker PA, Sicotte H, Pekmezci M, Rice T, Kosel ML, Smirnov IV, Sarkar G, Caron AA, Kollmeyer TM, Praska CE, Chada AR, Halder C, Hansen HM, McCoy LS, Bracci PM, Marshall R, Zheng S, Reis GF, Pico AR, O'Neill BP, Buckner JC, Giannini C, Huse JT, Perry A, Tihan T, Berger MS, Chang SM, Prados MD, Wiemels J, Wiencke JK, Wrensch MR, Jenkins RB (2015) Glioma groups based on 1p/19q, IDH, and TERT promoter mutations in tumors. N Engl J Med 372(26):2499–2508. https://doi.org/10.1056/NEJMoa1407279

19. Brat DJ, Verhaak RG, Aldape KD, Yung WK, Salama SR, Cooper LA, Rheinbay E, Miller CR, Vitucci M, Morozova O, Robertson AG, Noushmehr H, Laird PW, Cherniack AD, Akbani R, Huse JT, Ciriello G, Poisson LM, Barnholtz-Sloan JS, Berger MS, Brennan C, Colen RR, Colman H, Flanders AE, Giannini C, Grifford M, Iavarone A, Jain R, Joseph I, Kim J, Kasaian K, Mikkelsen T, Murray BA, O'Neill BP, Pachter L, Parsons DW, Sougnez C, Sulman EP, Vandenberg SR, Van Meir EG, von Deimling A, Zhang H, Crain D, Lau K, Mallery D, Morris S, Paulauskis J, Penny R, Shelton T, Sherman M, Yena P, Black A, Bowen J, Dicostanzo K, Gastier-Foster J, Leraas KM, Lichtenberg TM, Pierson CR, Ramirez NC, Taylor C, Weaver S, Wise L, Zmuda E, Davidsen T, Demchok JA, Eley G, Ferguson ML, Hutter CM, Mills Shaw KR, Ozenberger BA, Sheth M, Sofia HJ, Tarnuzzer R, Wang Z, Yang L, Zenklusen JC, Ayala B, Baboud J, Chudamani S, Jensen MA, Liu J, Pihl T, Raman R, Wan Y, Wu Y, Ally A, Auman JT, Balasundaram M, Balu S, Baylin SB, Beroukhim R, Bootwalla MS, Bowlby R, Bristow CA, Brooks D, Butterfield Y, Carlsen R, Carter S, Chin L, Chu A, Chuah E, Cibulskis K, Clarke A, Coetzee SG, Dhalla N, Fennell T, Fisher S, Gabriel S, Getz G, Gibbs R, Guin R, Hadjipanayis A, Hayes DN, Hinoue T, Hoadley K, Holt RA, Hoyle AP, Jefferys SR, Jones S, Jones CD, Kucherlapati R, Lai PH, Lander E, Lee S, Lichtenstein L, Ma Y, Maglinte DT, Mahadeshwar HS, Marra MA, Mayo M, Meng S, Meyerson ML, Mieczkowski PA, Moore RA, Mose LE, Mungall AJ, Pantazi A, Parfenov M, Park PJ, Parker JS, Perou CM, Protopopov A, Ren X, Roach J, Sabedot TS, Schein J, Schumacher SE, Seidman JG, Seth S, Shen H, Simons JV, Sipahimalani P, Soloway MG, Song X, Sun H, Tabak B, Tam A, Tan D, Tang J, Thiessen N, Triche T Jr, Van Den Berg DJ, Veluvolu U, Waring S, Weisenberger DJ, Wilkerson MD, Wong T, Wu J, Xi L, Xu AW, Yang L, Zack TI, Zhang J, Aksoy BA, Arachchi H, Benz C, Bernard B, Carlin D, Cho J, DiCara D, Frazer S, Fuller GN, Gao J, Gehlenborg N, Haussler D, Heiman DI, Iype L, Jacobsen A, Ju Z, Katzman S, Kim H, Knijnenburg T, Kreisberg RB, Lawrence MS, Lee W, Leinonen K, Lin P, Ling S, Liu W, Liu Y, Liu Y, Lu Y, Mills G, Ng S, Noble MS, Paull E, Rao A, Reynolds S, Saksena G, Sanborn Z, Sander C, Schultz N, Senbabaoglu Y, Shen R, Shmulevich I, Sinha R, Stuart J, Sumer SO, Sun Y, Tasman N, Taylor BS, Voet D, Weinhold N, Weinstein JN, Yang D, Yoshihara K, Zheng S, Zhang W, Zou L, Abel T, Sadeghi S, Cohen ML, Eschbacher J, Hattab EM, Raghunathan A, Schniederjan MJ, Aziz D, Barnett G, Barrett W, Bigner DD, Boice L, Brewer C, Calatozzolo C, Campos B, Carlotti CG Jr, Chan TA, Cuppini L, Curley E,

Cuzzubbo S, Devine K, DiMeco F, Duell R, Elder JB, Fehrenbach A, Finocchiaro G, Friedman W, Fulop J, Gardner J, Hermes B, Herold-Mende C, Jungk C, Kendler A, Lehman NL, Lipp E, Liu O, Mandt R, McGraw M, McLendon R, McPherson C, Neder L, Nguyen P, Noss A, Nunziata R, Ostrom QT, Palmer C, Perin A, Pollo B, Potapov A, Potapova O, Rathmell WK, Rotin D, Scarpace L, Schilero C, Senecal K, Shimmel K, Shurkhay V, Sifri S, Singh R, Sloan AE, Smolenski K, Staugaitis SM, Steele R, Thorne L, Tirapelli DP, Unterberg A, Vallurupalli M, Wang Y, Warnick R, Williams F, Wolinsky Y, Bell S, Rosenberg M, Stewart C, Huang F, Grimsby JL, Radenbaugh AJ, Zhang J (2015) Comprehensive, integrative genomic analysis of diffuse lower-grade gliomas. N Engl J Med 372(26):2481–2498. https://doi.org/10.1056/NEJMoa1402121

20. Brennan CW, Verhaak RG, McKenna A, Campos B, Noushmehr H, Salama SR, Zheng S, Chakravarty D, Sanborn JZ, Berman SH, Beroukhim R, Bernard B, Wu CJ, Genovese G, Shmulevich I, Barnholtz-Sloan J, Zou L, Vegesna R, Shukla SA, Ciriello G, Yung WK, Zhang W, Sougnez C, Mikkelsen T, Aldape K, Bigner DD, Van Meir EG, Prados M, Sloan A, Black KL, Eschbacher J, Finocchiaro G, Friedman W, Andrews DW, Guha A, Iacocca M, O'Neill BP, Foltz G, Myers J, Weisenberger DJ, Penny R, Kucherlapati R, Perou CM, Hayes DN, Gibbs R, Marra M, Mills GB, Lander E, Spellman P, Wilson R, Sander C, Weinstein J, Meyerson M, Gabriel S, Laird PW, Haussler D, Getz G, Chin L (2013) The somatic genomic landscape of glioblastoma. Cell 155 (2):462–477. https://doi.org/10.1016/j.cell.2013.09.034

21. Verhaak RGW, Hoadley KA, Purdom E, Wang V, Qi Y, Wilkerson MD, Miller CR, Ding L, Golub T, Mesirov JP, Alexe G, Lawrence M, O'Kelly M, Tamayo P, Weir BA, Gabriel S, Winckler W, Gupta S, Jakkula L, Feiler HS, Hodgson JG, James CD, Sarkaria JN, Brennan C, Kahn A, Spellman PT, Wilson RK, Speed TP, Gray JW, Meyerson M, Getz G, Perou CM, Hayes DN, Network CGAR (2010) Integrated genomic analysis identifies clinically relevant subtypes of glioblastoma characterized by abnormalities in PDGFRA, IDH1, EGFR, and NF1. Cancer Cell 17 (1):98–110. https://doi.org/10.1016/j.ccr.2009.12.020

22. Picard D, Miller S, Hawkins CE, Bouffet E, Rogers HA, Chan TS, Kim SK, Ra YS, Fangusaro J, Korshunov A, Toledano H, Nakamura H, Hayden JT, Chan J, Lafay-Cousin L, Hu P, Fan X, Muraszko KM, Pomeroy SL, Lau CC, Ng HK, Jones C, Van Meter T, Clifford SC, Eberhart C, Gajjar A, Pfister SM, Grundy RG, Huang A (2012) Markers of survival and metastatic potential in childhood CNS primitive neuro-ectodermal brain tumours: an integrative genomic analysis. Lancet Oncol 13(8):838–848. https://doi.org/10.1016/s1470-2045(12)70257-7

23. Torchia J, Picard D, Lafay-Cousin L, Hawkins CE, Kim SK, Letourneau L, Ra YS, Ho KC, Chan TS, Sin-Chan P, Dunham CP, Yip S, Ng HK, Lu JQ, Albrecht S, Pimentel J, Chan JA, Somers GR, Zielenska M, Faria CC, Roque L, Baskin B, Birks D, Foreman N, Strother D, Klekner A, Garami M, Hauser P, Hortobagyi T, Bognar L, Wilson B, Hukin J, Carret AS, Van Meter TE, Nakamura H, Toledano H, Fried I, Fults D, Wataya T, Fryer C, Eisenstat DD, Scheineman K, Johnston D, Michaud J, Zelcer S, Hammond R, Ramsay DA, Fleming AJ, Lulla RR, Fangusaro JR, Sirachainan N, Larbcharoensub N, Hongeng S, Barakzai MA, Montpetit A, Stephens D, Grundy RG, Schuller U, Nicolaides T, Tihan T, Phillips J, Taylor MD, Rutka JT, Dirks P, Bader GD, Warmuth-Metz M, Rutkowski S, Pietsch T, Judkins AR, Jabado N, Bouffet E, Huang A (2015) Molecular subgroups of atypical teratoid rhabdoid tumours in children: an integrated genomic and clinicopathological analysis. Lancet Oncol 16(5):569–582. https://doi.org/10.1016/s1470-2045(15)70114-2

24. Witt H, Mack SC, Ryzhova M, Bender S, Sill M, Isserlin R, Benner A, Hielscher T, Milde T, Remke M, Jones DT, Northcott PA, Garzia L, Bertrand KC, Wittmann A, Yao Y, Roberts SS, Massimi L, Van Meter T, Weiss WA, Gupta N, Grajkowska W, Lach B, Cho YJ, von Deimling A, Kulozik AE, Witt O, Bader GD, Hawkins CE, Tabori U, Guha A, Rutka JT, Lichter P, Korshunov A, Taylor MD, Pfister SM (2011) Delineation of two clinically and molecularly distinct subgroups of posterior fossa ependymoma. Cancer Cell 20 (2):143–157. https://doi.org/10.1016/j.ccr.2011.07.007

25. Taylor MD, Poppleton H, Fuller C, Su X, Liu Y, Jensen P, Magdaleno S, Dalton J, Calabrese C, Board J, Macdonald T, Rutka J, Guha A, Gajjar A, Curran T, Gilbertson RJ (2005) Radial glia cells are candidate stem cells of ependymoma. Cancer Cell 8 (4):323–335. https://doi.org/10.1016/j.ccr.2005.09.001

26. Merino DM, Shlien A, Villani A, Pienkowska M, Mack S, Ramaswamy V,

Shih D, Tatevossian R, Novokmet A, Choufani S, Dvir R, Ben-Arush M, Harris BT, Hwang EI, Lulla R, Pfister SM, Achatz MI, Jabado N, Finlay JL, Weksberg R, Bouffet E, Hawkins C, Taylor MD, Tabori U, Ellison DW, Gilbertson RJ, Malkin D (2015) Molecular characterization of choroid plexus tumors reveals novel clinically relevant subgroups. Clin Cancer Res 21(1):184–192. https://doi. org/10.1158/1078-0432.ccr-14-1324

27. Wang B, Mezlini AM, Demir F, Fiume M, Tu Z, Brudno M, Haibe-Kains B, Goldenberg A (2014) Similarity network fusion for aggregating data types on a genomic scale. Nat Methods 11(3):333–337. https://doi.org/ 10.1038/nmeth.2810

28. Cavalli FMG, Remke M, Rampasek L, Peacock J, Shih DJH, Luu B, Garzia L, Torchia J, Nor C, Morrissy AS, Agnihotri S, Thompson YY, Kuzan-Fischer CM, Farooq H, Isaev K, Daniels C, Cho BK, Kim SK, Wang KC, Lee JY, Grajkowska WA, Perek-Polnik M, Vasiljevic A, Faure-Conter C, Jouvet A, Giannini C, Nageswara Rao AA, KKW L, Ng HK, Eberhart CG, Pollack IF, Hamilton RL, Gillespie GY, Olson JM, Leary S, Weiss WA, Lach B, Chambless LB, Thompson RC, Cooper MK, Vibhakar R, Hauser P, van Veelen MC, Kros JM, French PJ, Ra YS, Kumabe T, Lopez-Aguilar E, Zitterbart K, Sterba J, Finocchiaro G, Massimino M, Van Meir EG, Osuka S, Shofuda T, Klekner A, Zollo M, Leonard JR, Rubin JB, Jabado N, Albrecht S, Mora J, Van Meter TE, Jung S, Moore AS, Hallahan AR, Chan JA, DPC T, Carlotti CG, Fouladi M, Pimentel J, Faria CC, Saad AG, Massimi L, Liau LM, Wheeler H, Nakamura H, Elbabaa SK, Perezpena-Diazconti M, Chico Ponce de Leon F, Robinson S, Zapotocky M, Lassaletta A, Huang A, Hawkins CE, Tabori U, Bouffet E, Bartels U, Dirks PB, Rutka JT, Bader GD, Reimand J, Goldenberg A, Ramaswamy V, Taylor MD (2017) Intertumoral heterogeneity within medulloblastoma subgroups. Cancer Cell 31(6):737–754.e736. https://doi.org/ 10.1016/j.ccell.2017.05.005

29. Capper D, Jones DTW, Sill M, Hovestadt V, Schrimpf D, Sturm D, Koelsche C, Sahm F, Chavez L, Reuss DE, Kratz A, Wefers AK, Huang K, Pajtler KW, Schweizer L, Stichel D, Olar A, Engel NW, Lindenberg K, Harter PN, Braczynski AK, Plate KH, Dohmen H, Garvalov BK, Coras R, Hölsken A, Hewer E, Bewerunge-Hudler M, Schick M, Fischer R, Beschorner R, Schittenhelm J, Staszewski O, Wani K, Varlet P, Pages M, Temming P, Lohmann D, Selt F, Witt H, Milde T, Witt O, Aronica E, Giangaspero F, Rushing E, Scheurlen W, Geisenberger C, Rodriguez FJ, Becker A, Preusser M, Haberler C, Bjerkvig R, Cryan J, Farrell M, Deckert M, Hench J, Frank S, Serrano J, Kannan K, Tsirigos A, Brück W, Hofer S, Brehmer S, Seiz-Rosenhagen M, Hänggi D, Hans V, Rozsnoki S, Hansford JR, Kohlhof P, Kristensen BW, Lechner M, Lopes B, Mawrin C, Ketter R, Kulozik A, Khatib Z, Heppner F, Koch A, Jouvet A, Keohane C, Mühleisen H, Mueller W, Pohl U, Prinz M, Benner A, Zapatka M, Gottardo NG, Driever PH, Kramm CM, Müller HL, Rutkowski S, von Hoff K, Frühwald MC, Gnekow A, Fleischhack G, Tippelt S, Calaminus G, Monoranu C-M, Perry A, Jones C, Jacques TS, Radlwimmer B, Gessi M, Pietsch T, Schramm J, Schackert G, Westphal M, Reifenberger G, Wesseling P, Weller M, Collins VP, Blümcke I, Bendszus M, Debus J, Huang A, Jabado N, Northcott PA, Paulus W, Gajjar A, Robinson GW, Taylor MD, Jaunmuktane Z, Ryzhova M, Platten M, Unterberg A, Wick W, Karajannis MA, Mittelbronn M, Acker T, Hartmann C, Aldape K, Schüller U, Buslei R, Lichter P, Kool M, Herold-Mende C, Ellison DW, Hasselblatt M, Snuderl M, Brandner S, Korshunov A, von Deimling A, Pfister SM (2018) DNA methylation-based classification of central nervous system tumours. Nature 555 (7697):469–474

30. Cavalli FMG, Hübner J-M, Sharma T, Luu B, Sill M, Zapotocky M, Mack SC, Witt H, Lin T, Shih DJH, Ho B, Santi M, Emery L, Hukin J, Dunham C, McLendon RE, Lipp ES, Gururangan S, Grossbach A, French P, Kros JM, van Veelen M-LC, Rao AAN, Giannini C, Leary S, Jung S, Faria CC, Mora J, Schüller U, Alonso MM, Chan JA, Klekner A, Chambless LB, Hwang EI, Massimino M, Eberhart CG, Karajannis MA, Benjamin L, Liau LM, Zollo M, Ferrucci V, Carlotti C, Tirapelli DPC, Tabori U, Bouffet E, Ryzhova M, Ellison DW, Merchant TE, Gilbert MR, Armstrong TS, Korshunov A, Pfister SM, Taylor MD, Aldape K, Pajtler KW, Kool M, Ramaswamy V (2018) Heterogeneity within the PF-EPN-B ependymoma subgroup. Acta Neuropathol 136 (2):227–237

Chapter 5

Detecting Stem Cell Marker Expression Using the NanoString nCounter System

Scott Ryall, Anthony Arnoldo, Javal Sheth, Sheila K. Singh, and Cynthia Hawkins

Abstract

The use of embryonic stem cells (ESCs) and induced pluripotent stem cells (iPSCs) has become common-place in the study of neuronal development, physiology, disease modelling, and therapy development. Due to the transient nature of working with these cells, it is important to regularly confirm the cell status as a naive stem cell versus a more defined neural progenitor cell (NPC). Classically, this has been done using a panel of specific antibodies to test for the expression of transcription factors known to be observed in ESCs, but not NPCs. However, this method is both time consuming and expensive. Here, we describe the use of the NanoString nCounter system for determining the levels of expression of key transcription factors that will effectively aid in determining the state of your stem cell cultures.

Key words Embryonic stem cells (ESCs), Induced-pluripotent stem cells (iPSCs), Neural, Neural progenitor, Transcription factor, Expression, NanoString nCounter

1 Introduction

Embryonic stem cells (ESCs) are defined as such by their pluripotent state and the ability to self-renew. As such, ESCs have the capacity to generate all the cell lineages of an adult organism while maintaining their ability to proliferate in the same state. ESCs possess a unique transcription factor expression profile that allows for these properties to be sustained, including a unique ground state and transcriptional circuitry [1–5]. These critical transcription factors were first identified in 2006 and enabled the generation of induced-pluripotent stem cells (iPSCs), stem cells that had been reprogrammed to a naive state [6]. Since their discovery, the use of ESC and iPSCs in research has revolutionized experimental mammalian genetics, development, physiology, disease modeling, and therapeutic development (Examples: [7–10]). As such, they remain one of the most frequently used in vitro models for scientists across many disciplines, including neuronal development, physiology, and disease.

Sheila K. Singh and Chitra Venugopal (eds.), *Brain Tumor Stem Cells: Methods and Protocols*, Methods in Molecular Biology, vol. 1869, https://doi.org/10.1007/978-1-4939-8805-1_5, © Springer Science+Business Media, LLC, part of Springer Nature 2019

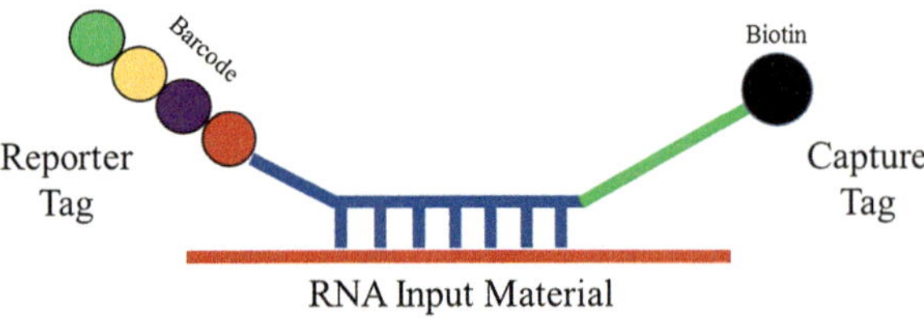

Fig. 1 Schematic of NanoString nCounter system. The unique barcode on the user defined binding target hybridizes to cellular RNA provided. The capture probe containing a biotin label at the 3′ end allows for isolation of the hybridized barcode

Due to the magnitude of research dedicated to understanding the gene expression of ESCs and iPSCs, key transcription factors attributed to a cell's "stemness" have been identified [6]. Due to the transient nature of working with naive stem cells [11], it is important that cells be regularly tested for these markers to differentiate them from neural progenitor cells (NPCs), a cell further differentiated along the neuronal developmental path. Most commonly, expression of OCT3/4, NANOG, and SOX2 are considered markers for naive ESC/iPSCs [12–15] while expression of NESTIN, PAX6, and SOX1 are associated with neural progenitor cells [16–18].

Currently, commercial antibody panels are available for the testing of these markers on cultured ESCs/iPSCs. However, this method is both time-consuming and expensive. Here we propose the use of the NanoString nCounter system for testing the level of expression of the above-mentioned stem cell markers [19]. The basic principles of this technique are shown in Fig. 1. This platform is both rapid and cost-effective and is capable of multiplexing additional targets of interest for your stem cell work. Although here we describe a protocol explicitly for neuronal stem cell work, the approach can easily be adapted to monitor the cell fate of diverse stem cell lineages using different transcripts.

2 Materials

All NanoString-specific materials and equipment used are commercially available for purchase from NanoString technologies. It is recommended that all materials be stored according to the manufacturer's suggestions and all waste be disposed of according to your institution's specifications. Note that common lab equipment (centrifuge, pipettes, balances, etc.) are not included in this section, but assumed to be present.

2.1 RNA Extraction

1. 2.0 mL prefilled bead tubes.

2. Direct-zol RNA MiniPrep Kit (Zymo Research R2050).

   (a) RNA Prewash.

   (b) RNA Wash Buffer (*see* **Note 1**).

   (c) DNase I.

   (d) DNA Digestion Buffer.

   (e) Zymo-Spin IIC Columns.

3. TRIzol Reagent (Thermo Scientific 15596026).

4. 95–100% Ethanol.

2.2 Probe Preparation

1. RNase-free water.

2. TE-Tween Buffer (10 nM Tris–HCL [pH 8.0], 1 mM EDTA, 0.1% Tween).

3. 1.7 mL microfuge tube.

4. Custom NanoString Probe sets (Table 1) (*see* **Note 2**).

2.3 Probe Hybridization

1. NanoString Hybridization Buffer TagSet.

2. 30× Working Probe A Pool (made in Subheading 3.2).

3. 30× Working Probe B Pool (made in Subheading 3.2).

4. RNase-free water.

5. PCR strip tubes.

Table 1
Suggested gene expression targets for distinguishing between ESCs/ iPSCs and NPC

Gene target	ESCs	NPCs	House-keeping
OCT3	×		
OCT4	×		
SOX2	×		
NANOG	×		
c-MYC	×		
NESTIN		×	
Pax6		×	
SOX1		×	
CREBBP [20]			×
PTEN [20]			×
RND1 [20]			×
MLH3 [20]			×

2.4 NanoString Expression Analysis	1. nCounter Prep Station. 2. nCounter Tips. 3. nCounter Tip Sheaths. 4. 12-Well Strip Tubes. 5. nCounter Cartridge Adhesive Cover. 6. 96-well preparation plates. 7. NanoString cartridge.
2.5 Data Analysis	1. USB key for data transfer.

3 Methods

3.1 RNA Extraction

1. Collect 1×10^6 of your cells of interest and place into a 2 mL bead tube (*see* **Note 3**).

2. Pellet the cells and remove any additional media from the bead tube.

3. Add 500 µL of TRIzol Reagent.

4. Lyse and homogenize the cells using a benchtop agitator.

5. Centrifuge the mixture at $12,000 \times g$ for 1 min (*see* **Note 4**).

6. Carefully transfer the supernatant into an RNase-free tube.

7. Discard the pellet.

8. Add 500 µL of 95–100% ethanol into the supernatant from **step 5** (*see* **Note 5**).

9. Mix well by vortexing.

10. Load the mixture into a Zymo-Spin IIC Column + collection tube.

11. Centrifuge for 1 min at $10,000–16,000 \times g$.

12. Transfer the column into a new collection tube and discard the collection tube containing the flow-through.

13. Add 400 µL RNA Wash Buffer.

14. Centrifuge for 1 min at $10,000–16,000 \times g$ and discard the collection tube containing the flow-through.

15. Transfer the column into a new collection tube.

16. Prepare DNase I cocktail: • 5 µL of DNase I (6 U/µL) • 75 µL of DNA Digestion Buffer.

17. Add 80 µL of the DNase I cocktail prepared in **step 15** directly to the matrix of the Zymo-Spin IIC Column.

18. Incubate the column at room temperature (20–30 °C) for 15 min.

19. Add 400 µL Direct-zol RNA Prewash to the column.

20. Centrifuge for 1 min at 10,000–16,000 × g and discard the flow-through.

21. Add 700 μL RNA Wash Buffer to the column.

22. Centrifuge for 1 min at 10,000–16,000 × g and discard the flow-through.

23. Centrifuge for an additional 2 min at full speed to dry the column membrane.

24. Transfer the column into an RNase-free tube.

25. Add 30 μL of DNase/RNase-Free water directly to the column matrix.

26. Let it stand at room temperature for about 1 min.

27. Centrifuge at max speed for 1 min.

28. Check the quality and quantity of your RNA as per your lab's standard methods (*see* **Note 6**).

29. Aliquot the total RNA and store at −80 °C.

3.2 Probe Preparation

3.2.1 Master Probe A Stock (See **Note 7***)*

1. Pipet 5 μL of each Probe A (starting concentration of 1 μM) into a 1.7 mL microfuge tube (*see* **Note 8**).

2. Add TE to a final combined volume of 1 mL.

3. The final concentration of each probe A in the master stock will be 5 nM.

4. Store in aliquots at −20 °C or −80 °C as recommended by supplier.

3.2.2 Master Probe B Stock (See **Note 7***)*

1. Pipet 5 μL of each Probe B (starting concentration of 5 μM) into a 1.7 mL microfuge tube (*see* **Note 8**).

2. Add TE to a final combined volume of 1 mL.

3. The final concentration of each probe B in the master stock will be 25 nM.

4. Store in aliquots at −20 °C or −80 °C as recommended by the supplier.

3.2.3 30× Working Probe A Pool (See **Note 9***)*

1. Start with the Master Probe A Stock (made in Subheading 3.2.1), follow the 8.3-fold dilution outlined in Table 2 to generate a 30× Working Probe A Pool at the appropriate scale.

2. Mix well and spin to bring contents to the bottom of the tube. The final concentration of each Probe A in the working pool will be 0.6 nM.

3. The final concentration of each Probe A in the 30 μL hybridization assay will be 20 pM (*see* **Note 10**).

Table 2
Dilution table for proper probe dilutions

Number of assays	Aliquot of stock	TE-Tween (μL)	Final volume (μL)
12	4	29	33
24	4	29	33
36	5	37	42
48	7	51	58
60	8	59	67
72	10	73	83
84	11	81	92
96	13	95	108
108	15	110	125
120	16	117	133
132	18	132	150
144	19	139	158
156	21	154	175
168	23	169	192
180	24	176	200
192	26	191	217

*3.2.4 30× Working Probe B Pool (See **Note 9**)*

1. Start with the Master Probe B Stock (made in Subheading 3.2.2), follow the 8.3-fold dilution outlined in Table 2 to generate a 30× Working Probe B Pool at the appropriate scale.

2. Mix well and spin to bring contents to the bottom of the tube. The final concentration of each Probe B in the working pool will be 3 nM.

3. The final concentration of each Probe B in the 30 μL hybridization assay will be 100 pM (*see* **Note 10**).

3.3 Probe Hybridization

1. Remove one tube of the NanoString TagSet from the freezer and thaw it on ice. Invert several times to mix well, and briefly spin down the reagent at less than $100 \times g$.

2. Remove one tube of the 30× Working Probe A Pool (made in Subheading 3.2.3) and one tube of the 30× Working Probe B Pool (made in Subheading 3.2.4) from the freezer and thaw it on ice. Mix well by flicking the tube, and briefly spin down the reagent at less than $100 \times g$.

Table 3
NanoString TagSet master-mix components

Component	Per reaction (μL)	Per 13 reactions (μL)
Hybridization buffer	10	130
TagSet	5	65
30× Working Probe A Pool	1	13
30× Working Probe B Pool	1	13
Sample	13	
Total volume	30	

3. Take the Hybridization Buffer from the NanoString reagent box and leave on the bench.

4. Ensure that the thermal cycler is ready to execute the hybridization step (*see* **step 10**).

5. Label a 12-tube strip and cut it in half so it will fit in a picofuge.

6. Double check that the samples are in the right order. Add total RNA sample to each tube.

7. Create a partial master mix by adding reagents in the following order: 130 μL of hybridization buffer, 13 μL of the 30× Working Probe A Pool, and 13 μL of the 30× Working Probe B Pool to complete the master mix (Table 3) (*see* **Note 11**).

8. Mix the wells by flicking the tube and briefly spin down.

9. Immediately add 17 μL of master mix to each of the 12 tubes using a fresh pipette tip for each tube.

10. Program the thermal cycler to use a 30 μL volume and heated lid at 72 °C with the program: 20 h at 67 °C, ramp down to 4 °C.

11. Cap tubes tightly and mix the reagents by inverting the strip tubes several times. Ensure complete mixing.

12. Briefly spin down the hybridization reactions at less than $100 \times g$ and immediately place the strip tubes in the thermal cycler.

13. Incubate reactions for at least 16 h at 67 °C. Maximum hybridization time should not exceed 48 h. Ramp reactions down to 4 °C and process the following day (*see* **Note 12**).

3.4 NanoString Expression Analysis

1. Remove two NanoString Prep Plates from the fridge and one NanoString cartridge from the freezer.

2. Set out the sealed Prep Plates and cartridge for at least 10 min to warm to room temperature before using.

3. Centrifuge the two Prep Plates at 2000 × *g* for 2 min (*see* **Note 13**).

4. Turn on the nCounter Prep Station by pressing the power switch located on the back of the instrument.

5. Remove the PCR tubes from the thermal cycler (*see* methods Subheading 3.3) and briefly spin down at less than 1000 × *g*.

6. Load the tubes onto the nCounter Prep Station (*see* **Note 14**).

7. Select the "high sensitivity" protocol and indicate the number of samples being run.

8. Follow the on-screen prompts until the "start run" option appears.

9. Start the run.

10. The machine will progress through the following steps: validation deck layout and system processing.

11. Upon completion of system processing, hit the "next" button to prompt the "run successfully completed" command.

12. Remove the plate from the nCounter Prep Station (*see* **Note 15**).

13. Transfer the plate to the nCounter digital analyzer corresponding to the sample information you will enter in **steps 16–18**.

14. Turn on the nCounter digital analyzer and assign a Reporter Library File (RLF).

15. To begin the run, select "start counting" from the main menu and define the start position.

16. The "select cartridge definition mode" screen will next appear leading you to the "cartridge information" section allowing you to enter information regarding the run and the cartridge.

17. In the "cartridge information" screen, be sure to select "fov count" and select 555 fields of view (FOV).

18. Once complete sample information has been entered, select "done."

19. Shut the instrument door and click the "start" button to initiate the machine.

20. Upon completion, the stage will move into a position for the door to be opened and cartridges removed.

3.5 Data Analysis

1. Using an USB drive, transfer the RLF Zipped File (RCC.ZIP) from the NanoString nCounter Digital Analyzer to the computer on which the analysis will be performed.

2. Unzip the folder(s) containing your RCC data.

3. Extract raw count numbers for each sample using the free NanoString analysis software nSolver (*see* **Note 16**).

4. Visualization of the data may be completed to the operator's preference, focussing on marker expression changes between samples as compared to the stable control targets.

5. Gene expression of the targets may be used to infer the cell state of the run sample (*see* **Note 3**).

4   Notes

1. Ensure that the appropriate amount of 96–100% ethanol is added before use.

2. Table 1 lists the gene targets we advise be used for determining naive versus differentiated stem cells. However, due to the expandable nature of the NanoString platform, additional targets may be added if desired. To do this, contact NanoString technologies for probe design. Proper validation of the probe set must be conducted by the practicing lab to ensure no cross-reactivity is observed.

3. It is suggested that a differentiated cell pellet is also collected to provide a control comparison, as transcription factor expression tends to exist on a spectrum.

4. Ensure all residual cell debris is removed. You may increase the centrifugation time as needed.

5. Ensure that the ethanol added is in a 1:1 ratio to the TRIzol Reagent.

6. We advise using a standard NanoDrop spectrophotometer (Thermo Scientific) to determine the quantity and quality of the RNA. We recommend using RNA with a 260/280 value of 1.9–2.1.

7. Do not create a combined master stock containing Probe As and Probe Bs in the same tube. This generates elevated background noise and lowers assay sensitivity.

8. NanoString probes are based on two key elements; a capture probe that directly binds the genetic material and a reporter probe which is responsible for binding the fluorescent barcode (TagSet) with each marker (Fig. 1). These are shipped separately. Probe sets A and B represent the reporter and capture probes, respectively, and must be made into working stocks individually.

9. Do NOT create a combined 30× working pool containing Probe As and Probe Bs in the same tube. This generates elevated background noise and lowers assay sensitivity.

10. Due to the dilute concentrations of many 30× working pools, long-term storage and reuse of these pools is not recommended. A fresh dilution of the master stock should be made for subsequent assays.

11. Always add "Probe Set B" last.

12. Do not leave the reactions at 4 °C for more than 24 h or increased background may result.

13. Using a medium deceleration speed will prevent splattering within the plate well.

14. When loading the cartridge, gently press the electrode down until locked into position. Do not force the tubes into place if resistance is experienced.

15. The cartridge must ideally be removed and sealed within 10 min of the prep run being completed. The sealed cartridge may be stored for up to 2 weeks at 4 °C if protected from light.

16. Four house-keeping genes are run alongside the genes of interest to normalize expression level to RNA quality.

Acknowledgments

Funding support from the Canadian Institute of Health Research (CIHR) Graduate Scholarship (S.R.).

References

1. Rossant J (2008) Stem cells and early lineage development. Cell 132:527–531

2. Silva J, Smith A (2008) Capturing pluripotency. Cell 132:532–536

3. Chen X, Xu H, Yuan P et al (2008) Integration of external signaling pathways with the core transcriptional network in embryonic stem cells. Cell 133:1106–1117

4. Jaenisch R, Young R (2008) Stem cells, the molecular circuitry of pluripotency and nuclear reprogramming. Cell 132:567–582

5. Macarthur BD, Ma'ayan A, Lemischka IR (2009) Systems biology of stem cell fate and cellular reprogramming. Nat Rev Mol Cell Biol 10:672–681

6. Takahashi K, Yamanaka S (2006) Induction of pluripotent stem cells from mouse embryonic and adult fibroblast cultures by defined factors. Cell 126:663–676

7. Carcamo-Orive L, Hoffman GE, Cundiff P et al (2017) Analysis of transcriptional variability in a large human iPSC library reveals genetic and non-genetic determinants of heterogeneity. Cell Stem Cell 20(4):518–532

8. Cayo M, Mallanna S, Di Furio F et al (2017) A drug screen using human iPSC-derived hepatocyte-like cells reveals cardiac glycosides as a potential treatment for hypercholesterolemia. Cell Stem Cell 20(4):478–489

9. Sahakyan A, Kim R, Chronis C et al (2017) Human naïve pluripotent stem cells model X chromosome dampening and X inactivation. Cell Stem Cell 20(1):87–101

10. Vallot C, Patrat C, Collier AJ et al (2017) XACT noncoding RNA competes with XIST n the control of X chromosome activity during human early development. Cell Stem Cell 20(1):102–111

11. Weinberger L, Ayyash M, Novershtern N et al (2016) Dynamic stem cell states: naïve to primed pluripotency in rodents and humans. Nat Rev Mol Cell Biol 17(3):155–169

12. Boyer L, Lee T, Cole M et al (2005) Core transcriptional regulatory circuitry in human embryonic stem cells. Cell 122:947–956

13. Kim J, Chu J, Shen X et al (2008) An extended transcriptional network for pluripotency of embryonic stem cells. Cell 132:1049–1061

14. Avilion A, Nicolis SK, Pevny LH et al (2003) Multipotent cell lineages in early mouse development depend on SOX2 function. Genes Dev 17:126–140

15. Chambers I, Colby D, Robertson M et al (2003) Functional expression cloning of Nanog, a pluripotency sustaining factor in embryonic stem cells. Cell 113:643–655

16. Hendrickson M, Rao A, Demerdash O et al (2011) Expression of nestin by neural cells in the adult rat and human brain. PLoS One 6(4): e18535

17. Zhang X, Huang CT, Chen J et al (2010) Pax6 is a human neuroectoderm cell fate determinant. Cell Stem Cell 7(1):90–100

18. Zhao S, Nichols J, Smith AG et al (2004) SoxB transcription factors specify neuroectodermal lineage choice in ES cells. Mol Cell Neurosci 27:332–342

19. Geiss G, Bumgarner R, Birditt B et al (2008) Direct multiplexed measurement of gene expression with color-coded probes. Nat Biotechnol 26(3):317–325

20. Synnergren J, Giesler TL, Adak S et al (2007) Differentiating human embryonic stem cells express a unique housekeeping gene signature. Stem Cells 25(2):473–480

Chapter 6

Flow Cytometric Analysis of Brain Tumor Stem Cells

Minomi K. Subapanditha, Ashley A. Adile, Chitra Venugopal, and Sheila K. Singh

Abstract

As a useful biotechnology, flow cytometry has revolutionized the field of cell analysis through its dynamic system that employs fluidics, optics, and electronics. It was first used to analyze DNA, but is often used to determine biomarker expression, as well as to characterize and sort cells, in accordance with various parameters. A common application of flow cytometry is the identification and isolation of a distinct cancer cell population, known as cancer stem cells (CSCs). Various biomarkers have been used to elucidate this proportion of cells within the brain, termed brain tumor initiating cells (BTICs). Here, we discuss methodology to prepare BTICs for flow cytometric analysis that includes the expression of markers.

Key words Flow cytometry, Cell sorting, Cell analysis biomarkers

1 Introduction

Since its debut in the late 1960s, flow cytometry was originally used for DNA analysis, having developed into a reliable and valuable tool for cell sorting based on multidimensional parameters [1]. Even with diverse applications, its design has stayed true to the initial structure, with advancements in its three primary systems—fluidics, optics, and electronics [2]. Modern flow cytometers provide accurate analysis of various biophysical, optical, and fluorescent characteristics of cells or particles.

In regards to its mechanism of action, the hydraulics system produces a stream of fluid, allowing single cells in suspension to individually pass through a specific point where the light source, often a laser, intersects the cells [1, 3, 4]. The optics system collects the scattered light and fluorescence emissions of each cell though a series of filters and electronic detectors that identify distinct wavelengths [1, 2, 4]. This information is immediately sent to computer analysis software to graph the distribution of cells relative to various parameters. Remarkably, the rate of flow exceeds thousands of

Sheila K. Singh and Chitra Venugopal (eds.), *Brain Tumor Stem Cells: Methods and Protocols*, Methods in Molecular Biology, vol. 1869, https://doi.org/10.1007/978-1-4939-8805-1_6, © Springer Science+Business Media, LLC, part of Springer Nature 2019

particles per second, given the highly automated nature of this technology [5].

A major application of flow cytometry includes research in various cancer types, such as ovarian, prostate, breast, cancer in Barrett's esophagus, pancreatic, colon, acute myeloid leukemia, and brain [6–14]. It is through cell surface marker identification that this biotechnology allows for the elucidation of specific cancer cell populations. One particular population of interest is denoted as cancer stem cells (CSCs), which represent cancer cells with stem-like characteristics. Using flow cytometry and cell surface glycoprotein CD133, Singh et al. identified and isolated brain CSCs, or brain tumor initiating cells (BTICs) from the bulk tumor population [13, 14]. Other cell surface markers, including CD15 and CD271 [15–18], are commonly associated with BTICs. Our laboratory has continued to study BTICs in the context of medulloblastoma and glioblastoma, representing the most common malignant childhood and adult brain tumors respectively. In this chapter, we present cell analysis and sorting, as well as internal and external biomarker as it pertains to BTIC research.

2 Materials

1. 1× TrypLE™ Express Enzyme, Phenol Red.

2. Liberase™.

3. DNase 1 mg/mL.

4. 2 mM EDTA in 1× DPBS no calcium, no magnesium (2 mM PBS-EDTA).

5. 5 mL round-bottom polystyrene test tube with 35 μm cell strainer snap cap.

6. 0.4% Trypan Blue solution.

7. Countess® Cell Counting Chamber Slides.

8. Round-bottomed polypropylene tubes.

9. Kimwipes® disposable wiper.

10. 7-AAD viability dye.

11. LIVE/DEAD® Fixable Dead Cell Stain Kit.

    Component A—fluorescent reactive dye.
    Component B—DMSO Kimwipes® disposable wiper.

12. Kimwipes® disposable wiper.

13. BD Cytofix/Cytoperm Kit: BD Cytofix/Cytoperm Buffer™, 10× BD Perm/Wash™ (dilute to 1× with distilled water before use).

14. Flow cytometer (MoFlo™ XDP, Beckman Coulter).

3   Methods

3.1   Guide to Selecting Fluorophores

1. Determine availability of fluorophores conjugated to an antibody specific to marker(s) of study. Ensure excitation and emission spectra of fluorophore(s) correspond to laser and channel specifications of flow cytometer.

2. Correlate brightness of the fluorophore conjugate to the size of subpopulation when performing multiparameter analysis, such that an antibody detecting a rare population is conjugated to a bright fluorophore (e.g., phycoerythrin (PE)).

3. For multiparameter analysis, if possible, select fluorophores with minimal spectral overlap to avoid intensive compensation.

3.2   Preparing Adherent Cells for Staining

1. Dissociate cells into a single cell suspension by treating adherent cells with a sufficient volume of $1 \times$ TrypLE™ to submerge cells on the culture plate (*see* **Note 1**).

2. Incubate the plate at 37°C for 5 min, and wash with 5–10 mL of $1 \times$ DPBS to collect cell suspension into a conical centrifuge tube.

3. Centrifuge at $290 \times g$ for 5 min, and aspirate the supernatant.

4. Resuspend the pellet in 1 mL of 2 mM PBS-EDTA and filter the suspension through a 35 μm cell strainer in a 5 mL polystyrene test tube (*see* **Note 3**).

5. Use an automated cell counter with Countess® slides and 0.4% Trypan Blue solution to determine cell count and viability of filtered cell suspension (*see* **Note 4**).

3.3   Preparing Cells in Suspension for Staining

1. Resuspend pellet in 1 mL $1 \times$ DPBS and add 1 μg/μL Liberase™ to dissociate tumor spheres in cell culture (*see* **Note 2**).

2. Incubate at 37°C for 5 min or until tumor spheres dissociate.

3. Wash with approximately 1 mL $1 \times$ DPBS and centrifuge at $290 \times g$ for 5 min; aspirate the supernatant.

4. Resuspend pellet in 1 mL $1 \times$ DPBS and add 5–10 μg of DNase to rid of free-floating DNA in cell suspension.

5. Incubate at 37°C for 5 min, and wash with approximately 1 mL $1 \times$ DPBS and centrifuge at $290 \times g$ for 5 min; aspirate the supernatant.

6. Resuspend pellet in 1 mL of 2 mM PBS-EDTA and filter the suspension through a 35 μm cell strainer in a 5 mL polystyrene test tube (*see* **Note 3**).

7. Use an automated cell counter with Countess® slides and 0.4% Trypan Blue solution to determine cell count and viability of filtered cell suspension (*see* **Note 4**).

3.4 Extracellular Staining

1. Make required number of 100 µL aliquots of cell suspension in round-bottomed polypropylene tubes such that each aliquot will contain approximately 300,000 viable cells of prefiltered cell suspension.

2. Add select conjugated antibodies, or unconjugated primary antibodies with a paired conjugated secondary antibody, or matched isotype controls to tubes with cell aliquots at dilutions determined through antibody titration, and resuspend thoroughly (*see* **Notes 5** and **6**).

3. Incubate stained cell suspensions in the dark at 4°C for 15 min, or on ice for 30 min.

4. Wash cells with approximately 1 mL of 2 mM PBS-EDTA and centrifuge at 290 × *g* for 3 min.

5. Decant the supernatant into 20% bleach solution and blot with a Kimwipes® disposable wiper to avoid mixing of residual supernatant in tube with cell pellet. Confirm presence of cell pellet following centrifugation to avoid loss of cells during discard of supernatant.

6. Resuspend pellets thoroughly in 250 µL of 2 mM PBS-EDTA.

7. Add 2% 7-AAD and incubate at room temperature for 10 min, and analyze on flow cytometer (*see* **Note 7**).

3.5 Intracellular Staining

1. Add 1 µL of LIVE/DEAD® Fixable dye premixed with DMSO to a round-bottomed polypropylene tube containing approximately one million cells in 1 mL suspension, and incubate in the dark at 4°C for 20 min.

    (a) Select a viability dye that can be used for fixed cells, such as LIVE/DEAD® Fixable Dead Cell Stain, and has fluorescence compatible with parameters of the antibody conjugates.

2. Wash cells with approximately 1 mL of 2 mM PBS-EDTA and centrifuge at 290 × *g* for 3 min.

3. Decant the supernatant into 20% bleach solution and blot with a Kimwipes® disposable wiper to avoid mixing of residual supernatant in tube with cell pellet. Thoroughly resuspend the pellet by "flicking" the tube to avoid cell aggregation (*see* **Note 8** *before you proceed*).

4. Add 200 µL of BD Cytofix/Cytoperm Buffer™ from BD Cytofix/Cytoperm Kit to each tube to fix and permeabilize the cells, and incubate at 4°C for 20 min.

5. Wash cells with approximately 1 mL of 1× BD Perm/Wash™ buffer and centrifuge at 290 × *g* for 3 min.

6. Decant the supernatant into 20% bleach solution and blot with a Kimwipes® disposable wiper.

7. Make required number of 100 μL aliquots by resuspending thoroughly the fixed/permeabilized cells with 1× BD Perm/Wash™ buffer.

8. Add select conjugated antibodies, or unconjugated primary antibodies with a paired conjugated secondary antibody, or matched isotype controls to cell aliquots at dilutions determined through antibody titration and resuspend thoroughly (*see* **Notes 5** and **6**).

9. Incubate stained cell suspension in the dark at 4°C for 30 min.

10. Wash cells with approx. 1 mL of 1× BD Perm/Wash™ buffer and centrifuge at 290 × *g* for 3 min.

11. Decant the supernatant into 20% bleach solution and blot with a Kimwipes® disposable wiper.

12. Resuspend pellets thoroughly in 250 μL of 2 mM PBS-EDTA, and analyze on flow cytometer.

3.6 Setting Up for Analysis

1. Adjust forward scatter (size of cells) and side scatter (granularity of cells) using an unstained sample of cells, and set up a proper FSC vs. SSC plot for cells of study (*see* **Note 9**, Fig. 1a).

2. Use a positive control of cells stained with select dye to set region for viability as well as for compensation with fluorescent channels. Sample can be prepared by staining cells obtained from treatment with either heat or mild concentrations of bleach (*see* **Note 10**, Fig. 1b).

3. Use single stained controls to compensate for spectral overlap between fluorophores when multicolor analysis is performed. Compensation beads could replace cells that do not display optimal positive vs. negative levels of antibody binding for proper compensation (*see* **Note 11**, Fig. 2).

4. Use a negative control, or matched-isotype control, to adjust the background fluorescence for each parameter of select multicolor panel and ensure levels lie between the first two decades (10^0 to 10^2) (*see* **Note 12**, Fig. 3a).

5. Set gating strategies with unstained, negative or matched-isotype, and positive controls to obtain clear populations for test samples (Fig. 3b).

6. Run samples and collect at least 10,000 events for population of interest to ensure statistical significance.

3.7 Setting Up for Sorting

1. Follow **steps 1–5** of Subheading 3.6.

2. Set up required number of collection tubes for sort populations (number of sort streams is instrument-dependent) by coating 5 mL polypropylene tubes with collection buffer—either cell culture media, or a mixture of 2% FBS in 2 mM PBS-EDTA (*see* **Note 13**).

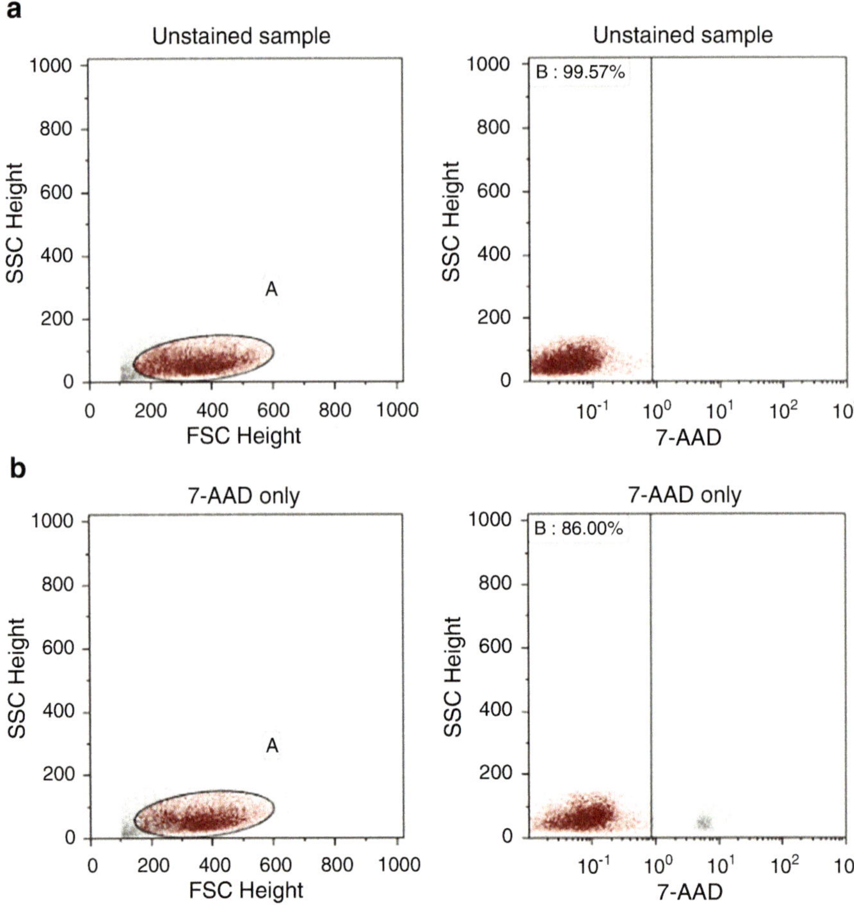

Fig. 1 Setting gates for viable cells. An unstained sample is used to set an appropriate SSC vs. FSC plot as well as region negative for dye staining (**a**), in this case 7-AAD. Viability dye 7-AAD binds to DNA, and therefore does not bind to viable cells with intact cell membranes, giving rise to a separation between viable and nonviable cells (**b**)

3. Decant enough collection buffer to leave approximately 500 μL in tube and set up tube to collect selected subpopulation.

4. Keep the collection tube(s) cold, especially if sorting large numbers of cells (This option may be instrument-dependent).

4   Notes

1. Add approx. 2 mL to a 100 mm × 20 mm culture dish; 1 mL to a 60 mm × 15 mm culture dish; and 500 μL to each well of a 6-well culture plate.

2. Adding liberase to cells grown in suspension as tumor spheres is a better alternative to treating cells just with TrypLE™

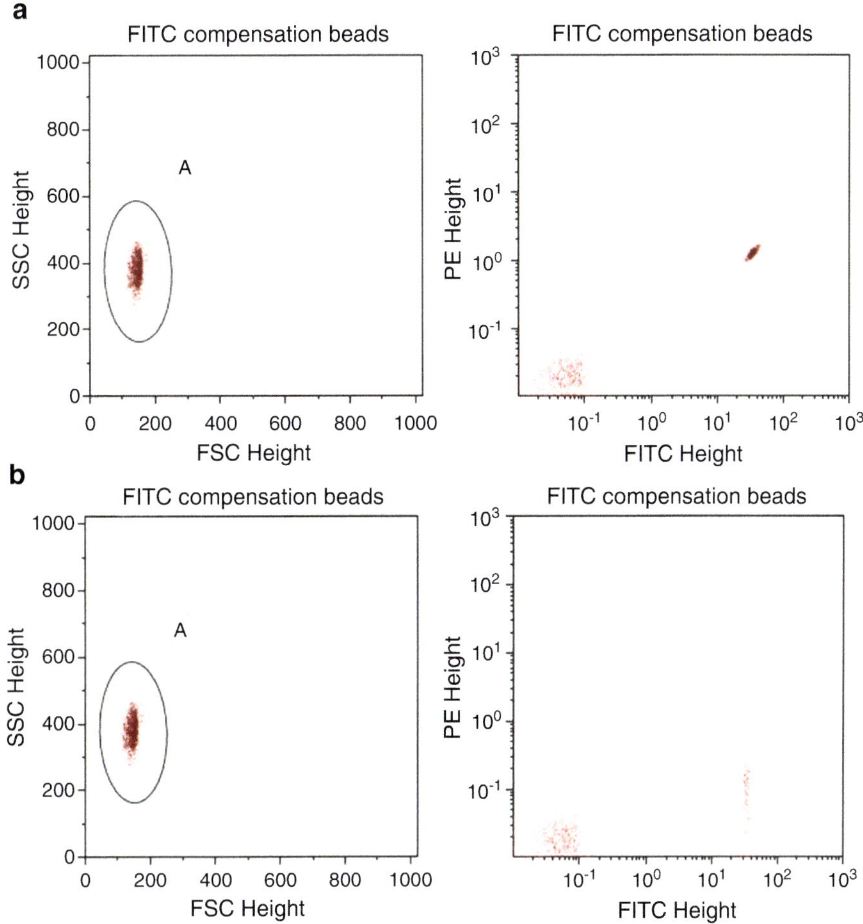

Fig. 2 Using compensation beads to compensate for spectral overlap between fluorophores. Compensation beads incubated with antibody conjugated to fluorescein isothiocyanate (FITC) are run against channel for phycoerythrin (PE) (**a**); spillover of FITC signal in PE channel is compensated (**b**)

3. EDTA is used for its chelating properties, which helps reduce clumping of cells caused by presence of ions. The concentration of EDTA may be increased if cells of study tend to be very clumpy.

4. A hemocytometer can be alternatively used to determine the cell count.

5. Cells can be stained with unconjugated primary antibodies as long as a suitable conjugated secondary antibody is used. Cells are incubated with primary antibody in predetermined concentration for 30 min at 4°C, washed with 1 mL of 2 mM PBS-EDTA, and incubated with select conjugated secondary antibody for 30 min at 4°C.

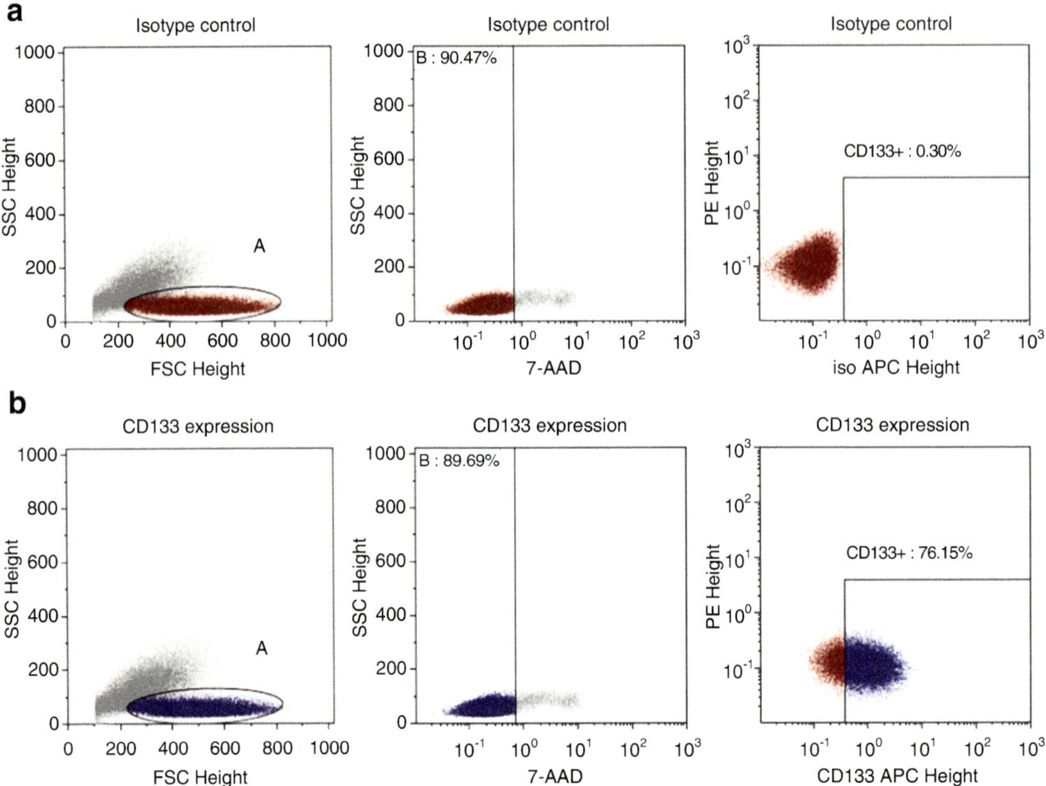

Fig. 3 Setting up gating strategies for protein expression. An isotype control is used to choose an appropriate region negating background fluorescence against a channel of choice (**a**), and sample with conjugated antibody of choice is run to characterize the cells for expression of protein (**b**)

6. Titration of antibodies for cells of study is recommended to obtain an optimal dilution of antibody where signal is maximized while background fluorescence is minimized. Set up a panel of increasing antibody dilutions and stain precounted cells along with negative and positive controls.

7. Presence of free-floating DNA, as a result of dead cells, also increases clumping of cells. The clumping effects of DNA can be reduced with addition of 2.5–5 ng of DNase.

8. Expression of surface antigens and intracellular proteins can be simultaneously analyzed by carrying out steps for staining of surface proteins prior to fixation/permeabilization of cells.

9. The FSC vs. SSC plot could be denoted either by FSC height vs. SSC height, or FSC area vs. SSC area.

10. For optimal staining with viability dye, use less harsh treatment conditions such that populations of both viable and nonviable cells are present.

11. Conjugated antibody is incubated with negative and positive compensation beads at the same dilution as added to the cells.

12. For single color analysis, the plot can be set in a way to present the channel for conjugated antibody against a "dump" channel, i.e., a channel with least spectral overlap with channel of antibody.

13. Use the same media to coat the collections as the media used for culturing of cells. For example, use Neurocult complete + NSA supplement (FGF, EGF, Heparin, antibiotics/antimycotics) for primary brain tumor samples. Add FBS to make up to 10%

References

1. Herzenberg LA, Parks D, Sahaf B et al (2002) The history and future of the fluorescence activated cell sorter and flow cytometry: a view from stanford. Clin Chem 48:1819–1827

2. Shapiro HM (2003) Practical flow cytometry. John Wiley & Sons, Hoboken, NJ

3. Brown M, Wittwer C (2000) Flow cytometry: principles and clinical applications in hematology. Clin Chem 46:1221–1229

4. Álvarez-Barrientios A, Arroyo J, Cantón R, Nombela C, Sánchez-Pérez M (2000) Applications of flow cytometry to clinical microbiology. Clin Microbiol Rev 13:167–195

5. Suthanthiraraj PPA, Graves S (2014) Fluidics. Curr Protoc Cytom:1–22

6. Friedlander ML, Hedley DW, Taylor IW et al (2006) Influence of cellular DMA content on survival in advanced ovarian cancer. Cancer Res 44:397–400

7. Collins AT, Berry PA, Hyde C et al (2005) Prospective identification of tumorigenic prostate cancer stem cells. Cancer Res 65:10946–10952

8. Dressler LG, Seamer LC, Owens MA, Clark GM, McGuire WL (1988) DNA flow cytometry and prognostic factors in 1331 frozen breast cancer specimens. Cancer 61:420–427

9. Reid BJ, Levine DS, Longton G et al (2000) Predictors of progression to cancer in Barrett's esophagus: baseline histology and flow cytometry identify low- and high-risk patient subsets. Am J Gastroenterol 95:1669–1676

10. Li C, Heidt DG, Dalerba P, Burant CF et al (2007) Identification of pancreatic cancer stem cells. Cancer Res 67:1030–1037

11. Brien CAO, Pollett A, Gallinger S, Dick JE (2007) A human colon cancer cell capable of initiating tumor growth in immunodeficient mice. Nat Lett 445:106–110

12. Bonnet D, Dick JE (1997) Human acute myeloid leukemia is organized as a hierarchy that originates from a primitive hematopoietic cell. Nat Med 3:730–737

13. Singh SK, Clarke ID, Terasaki M et al (2003) Identification of a cancer stem cell in human brain tumors. Cancer Res 63:5821–5828

14. Singh SK, Hawkins C, Clarke ID et al (2004) Identification of human brain tumour initiating cells. Nature 432:396–401

15. Read TA, Fogarty MP, Markant SL et al (2009) Identification of CD15 as a marker for tumor-propagating cells in a mouse model of medulloblastoma. Cancer Cell 15:135–147

16. Son MJ, Woodlard K, Nam DH et al (2009) SSEA-a is an enrichment marker for tumor-initiating cells in human glioblastoma. Cancer Stem Cell 4:440–452

17. Sonnenfeld KH, Ishii DN (1982) Nerve growth factor effects and receptors in cultured human neuroblastoma cell lines. J Neurosci Res 8:375–391

18. Barnes M, Eberhart CG, Collins R, Tihan T (2009) Expression of p7NTR in fetal brain and medulloblastmas: evidence of a precursor cell marker and its persistence in neoplasia. J Neuro-Oncol 92:193–201

Chapter 7

In Vitro Self-Renewal Assays for Brain Tumor Stem Cells

Mathieu Seyfrid, David Bobrowski, David Bakhshinyan, Nazanin Tatari, Chitra Venugopal, and Sheila K. Singh

Abstract

Early development of human organisms relies on stem cells, a population of non-specialized cells that can divide symmetrically to give rise to two identical daughter cells, or divide asymmetrically to produce one identical daughter cell and another more specialized cell. The capacity to undergo cellular divisions while maintaining an undifferentiated state is termed self-renewal and is responsible for the maintenance of stem cell populations during development. In addition, self-renewal plays a crucial role in the homeostasis of developed organism through replacement of defective cells.

Similar to their non-malignant counterparts, it has been postulated that tumor cells follow a differentiation hierarchy, with the least differentiated cells termed cancer stem cells (CSCs) at the apex. These tumor stem cells possess the ability to self-renew, have a higher capacity to initiate tumor growth when xenografted into an animal model, and can recapitulate the cell heterogeneity of the tumor they originate from. Hence, further investigation of mechanisms governing the self-renewal in cancer can lead to development of novel therapies targeting CSCs.

In this chapter, we described the soft agar assay and the limiting dilution assay (LDA) as two easy-to-implement and inexpensive assays to measure the stemness properties of brain tumor stem cells (BTSCs). These techniques constitute useful tools for the preclinical evaluation of therapeutic strategies targeting BTSCs clonogenicity.

Key words Brain tumor stem cells, Self-renewal assay, Anchorage-independent growth, Limiting dilution assay, Soft agar assay

1 Introduction

The establishment of colony formation assays started more than 60 years ago, for the purpose of studying clonogenicity both in normal and cancer cells [1, 2]. Following this pioneer work, in vivo experiments established a correlation between in vitro colony formation and in vivo self-renewal [3]. Decades later, Bonnet and Dick isolated cancer stem cells for the first time, from leukemia, by discriminating these cells with respect to their self-renewal status [4]. Following this discovery, tumor stem cells were then identified in solid tumors, such as breast [5] and brain [6], based on stemness properties

Sheila K. Singh and Chitra Venugopal (eds.), *Brain Tumor Stem Cells: Methods and Protocols*, Methods in Molecular Biology, vol. 1869, https://doi.org/10.1007/978-1-4939-8805-1_7, © Springer Science+Business Media, LLC, part of Springer Nature 2019

and markers. These breakthroughs highlighted the utility of self-renewal assays and established them as gold standard in cancer stem cells identification [7].

Under anchorage-free conditions, the majority of non-transformed cells undergo anoikis, a specific type of cell death provoked by loss of cell adhesion [8]. On the contrary, tumor stem cells can grow independently of a solid surface [9]. It has been shown that brain tumor stem cells (BTSCs) can be enriched in a serum-free, suspension culture environment [10]. Here, we describe in detail two in vitro techniques to evaluate the clonogenic potential of BTSCs as a measure of their self-renewal capacities. The soft agar formation assay is used to evaluate the clonogenicity of BTSCs in vitro by measuring colony formation in a semisolid matrix. These growth conditions mimic more closely the three dimensional in vivo environment and prevent spontaneous aggregation of spheres, thereby ensuring independent colony formation from single cells [11]. While still utilizing the self-renewal properties of BTSCs, the limiting dilution assay (LDA) is commonly used to determine the frequency of self-renewing cells in a cell culture. Both assays are effective tools for characterizing the stemness profile of putative BTSC populations, as a prelude to stemness markers monitoring and in vivo engraftment.

2 Materials

2.1 Propagation of BTSCs for Self-Renewal Assays

1. BTSC culture medium is NeuroCult™ complete media: NS-A Basal Medium (STEMCELL Technologies) with NeuroCult™NS-A Proliferation Supplement, EGF (20 ng/mL), FGF (10 ng/mL) and Heparin (2 µg/mL).
2. Liberase (0.2 Wünsch unit/mL) (Roche Applied Science).
3. DNase (1000 units/mL).
4. Trypan Blue stain 0.4%.
5. Lab-Line Aquabath.
6. 37°C incubator with 5% CO_2.
7. Countess II FL Cell Counter.
8. Centrifuge.
9. Microscope.
10. 15 mL Centrifuge Tube.
11. 50 mL Centrifuge Tube.
12. 5 mL Polystyrene round-bottom tube with cell-straining cap.
13. Brain tumor stem cell population.

2.2 Soft Agar Assay and LDA

1. UltraPure™ LMP (low melting point) Agarose (Invitrogen).
2. Eppendorf tubes.
3. Standard heating block.
4. Vortex mixer.
5. Aluminum foil.
6. Microwave.
7. 96-well plates.

3 Methods

3.1 Preparation of BTSCs

1. Transfer BTSCs cultured as tumorspheres in Neurocult complete media to a 15 mL tube and centrifuge them at $450 \times g$ for 5 min.

2. Dissociate BTSCs tumorspheres using adequate amount of Libarase and DNase (to avoid any cell-DNA clumping) depending on the cell pellet size. Incubate the tube at 37°C for 5–10 min.

3. Gently mix the solution using a 1 mL pipette to help the dissociation process. By adding 5 mL of PBS or Neurocult complete media, neutralize the effect of Librase and DNase. Centrifuge cells at $450 \times g$ for 5 min.

4. Remove the supernatant, resuspend cells in adequate amount of Neurocult complete media, and pass it through a 35 μm-pore filter to obtain a single cell suspension.

3.2 Soft Agar Assay (Fig. 1)

1. Weigh the appropriate quantity of Agarose and resuspend it in BTSC culture medium in order to obtain a 2.8% m/v Agarose solution (e.g., 28 mg LMP Agarose in 1 mL BTSC culture medium) (*see* **Note 1**).

2. Once homogenous, prepare a 1.4% Agarose solution by diluting the above solution in tissue culture media with a 1:1 ratio.

3. Count cells and prepare an appropriate cell solution. Seed at least triplicates for each condition (*see* **Note 2**).

4. Add adequate volume of 1.4% Agarose solution to each cell-containing well to reach a final concentration of 0.35% Agarose per well (*see* **Note 3**).

5. Return the plate to the incubator in order for the Agarose mixture to solidify.

6. Monitor the plate daily to determine when cells form visible colonies (*see* **Note 4**).

7. Once discernable colonies have formed, count them under 10× magnification. The number of independent colonies will be determined for each treatment condition, and compared to the control (*see* **Note 5**).

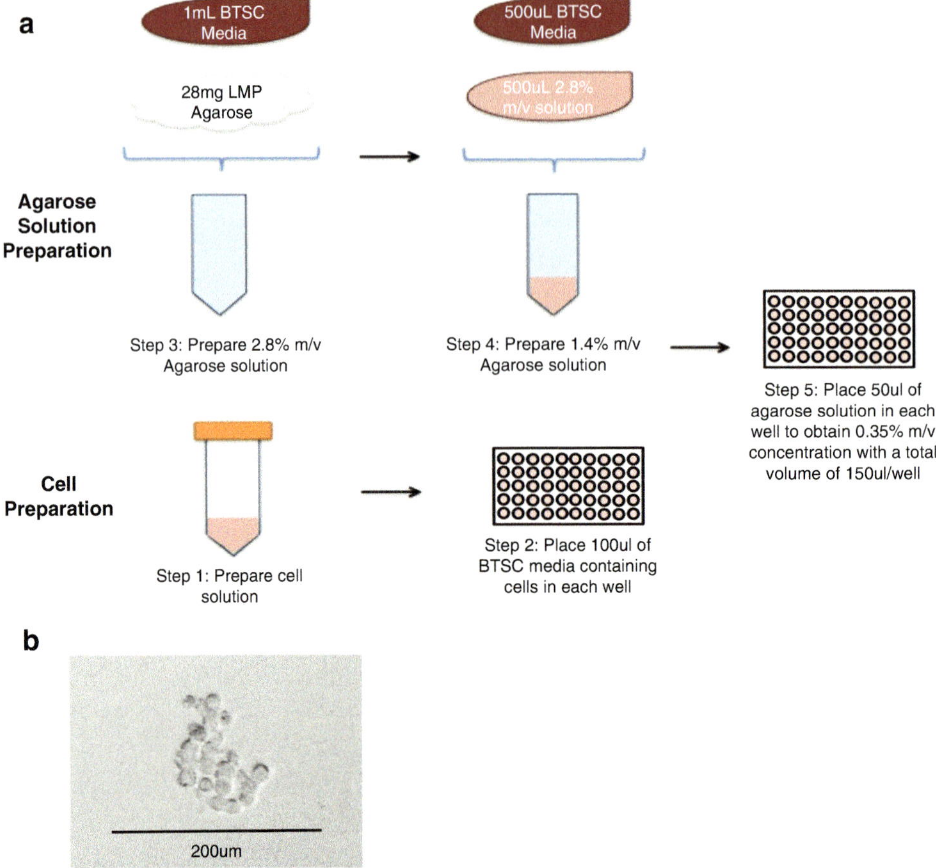

Fig. 1 Soft agar assay protocol and results. (**a**) Schematic representation of cell and Agarose solution preparations and plating for the soft agar protocol. (**b**) Colony formation from single cell in Agarose matrix following 7 days of incubation with 100 GBM cells seeded per well

3.3 Limiting Dilution Assay (LDA) (Fig. 2)

1. Prepare and seed a range of different cell concentrations (*see* **Notes 6** and **7**) in a 96-well plate in triplicates.

2. Return the plate to the incubator.

3. Count the number of wells per conditions that contain colonies, under 10× magnification (*see* **Note 4**).

4. The frequency of BTSCs within a given cell population is determined by linear regression analysis. Data is displayed as a scatter plot graph and the corresponding trend line; on the Y-axis the percentage of wells without detectable spheres and on the X-axis the number of seeded cells per well. Based on the Poisson distribution, the frequency of BTSCs in the sample is the value corresponding to 37% of wells without detectable spheres [12] (*see* **Note 8**).

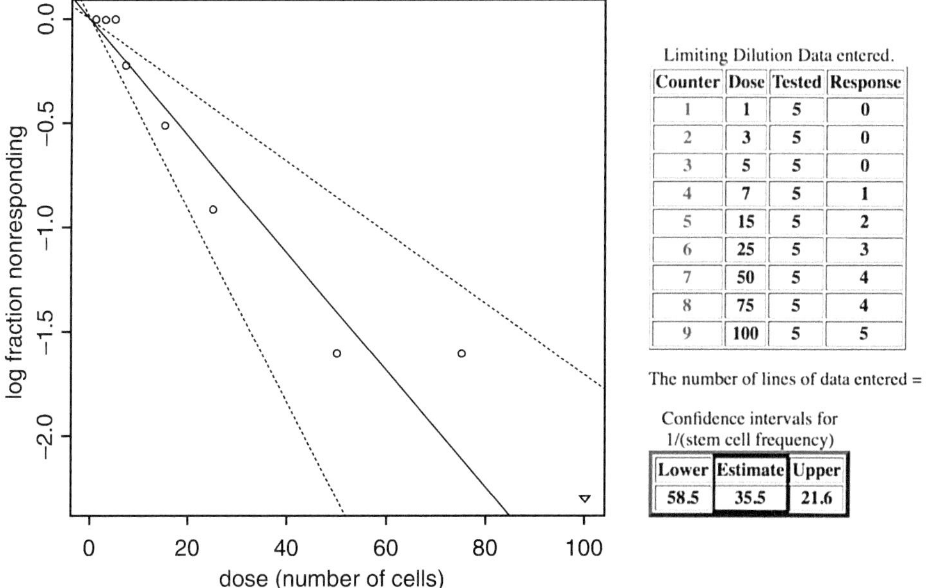

Fig. 2 Representative LDA regression curve plotted using the Extreme Limiting Dilution program (available from: http:/bioinf.wehi.au/software/elda/). The frequency of BTSCs in this representative GBM sample is the value corresponding to 37% of wells without detectable spheres; here, one BTSC every 36 cells

4 Notes

1. Intermitent vortexing will accelerate the solubilization process. According to the manufacturer's description, LMP Agarose melts at 65.5 °C, remains fluid above 37 °C, and will set rapidly at temperatures below 25 °C.

2. We sorted our cells using flow cytometry with an appropriate live-dead staining. Alternatively, count cells using a vital stain (e.g., trypan blue). To achieve a clear readout, it is optimal to have a cell solution with at least 85% viability. Different cell concentrations per well should be initially tested to assess which one is the most appropriate. In our case, 100 cells/ wells was the most consistent. It is important to note that for drug evaluation, any compounds to be tested should be used prior to seeding cells in the agar.

3. Place Agarose-media solution in a beaker filled with hot water to retain the Agarose solution fluidity. Be cautious with high temperatures, as they can be hazardous for cells in solution. Maintain the Agarose solution around 40°C, add it to the wells, and briefly but gently mix using a pipette for optimal results. Use caution to avoid introducing any air bubbles into the plate wells.

4. It is important to note that cell cycle kinetics influence the pace of tumorsphere formation. For this reason, regular monitoring of the seeded plate is required to determine the optimal experimental endpoint as distinct BTSC populations form spheres at different rates. In our case, visible colonies began to form after 7 days.

5. In addition, cell prestaining with an appropriate dye can facilitate counting colonies.

6. We used 250, 100, 75, 50, 25, 10, 7, 5, 3, and 1 cell per well. We advise seeding four or more replicates for a sensitive readout.

7. Count cells using a vital stain (e.g., trypan blue) and perform serial dilutions to obtain desired cell concentrations. Alternatively, cells can be sorted into a 96-well plate using flow cytometry with an appropriate live-dead dye.

8. Any appropriate scientific data analysis software, such as GraphPad Prism, can also be used. We used Extreme Limiting Dilution program (available from: http://bioinf.wehi.edu.au/software/elda/) to plot our data and to obtain the frequency of BTSCs in our sample.

References

1. Puck TT, Marcus PI, Cieciura SJ (1956) Clonal growth of mammalian cells in vitro; growth characteristics of colonies from single HeLa cells with and without a feeder layer. J Exp Med 103:273–283

2. Cieciura SJ, Marcus PI, Puck TT (1956) Clonal growth in vitro of epithelial cells from normal human tissues. J Exp Med 104:615–628

3. Becker AJ, McCulloch EA, Till JE (1963) Cytological demonstration of the clonal nature of spleen colonies derived from transplanted mouse marrow cells. Nature 197:452–454

4. Bonnet D, Dick JE (1997) Human acute myeloid leukemia is organized as a hierarchy that originates from a primitive hematopoietic cell. Nat Med 3:730–737

5. Al-Hajj M, Wicha MS, Benito-Hernandez A et al. (2003) Prospective identification of tumorigenic breast cancer cells. Proc Natl Acad Sci U S A 100:3983–3988

6. Singh SK, Hawkins C, Clarke ID et al. (2004) Identification of human brain tumour initiating cells. Nature 432:396–401

7. Reya T, Morrison SJ, Clarke MF, Weissman IL (2001) Stem cells, cancer, and cancer stem cells. Nature 414:105–111

8. Taddei ML, Giannoni E, Fiaschi T, Chiarugi P (2012) Anoikis: an emerging hallmark in health and diseases. J Pathol 226:380–393

9. Selby P, Buick RN, Tannock I (1983) A critical appraisal of the "human tumor stem-cell assay". N Engl J Med 308:129–134

10. Venugopal C, McFarlane NM, Nolte S et al. (2012) Processing of primary brain tumor tissue for stem cell assays and flow sorting. J Vis Exp (67)

11. Abbott A (2003) Cell culture: biology's new dimension. Nature 424:870–872

12. Hu Y, Smyth GK (2009) ELDA: extreme limiting dilution analysis for comparing depleted and enriched populations in stem cell and other assays. J Immunol Methods 347:70–78

Chapter 8

Differentiation of Brain Tumor Initiating Cells

Michelle M. Kameda-Smith, Minomi K. Subapanditha, Sabra K. Salim, Chitra Venugopal, and Sheila K. Singh

Abstract

Differentiation is a central key capability of stem cells. Their ability to be multipotent and undergo self-renewal are key identifying features of stem cells. A differentiation assay allows for study of one of the essential features of stem cells, the ability to differentiate into all of the cell types of its lineage, in order to ensure that the cells cultured and utilized in key experiments indeed have stem cell properties. Neural stem cells when plated in differentiation media, differentiate into all three neural lineages: Neurons, Astrocytes, and Oligodendrocytes. Brain tumor initiating cells (BTICs) are cells present in brain tumors that possess stem cell properties and are able to self-renew and differentiate into neural lineages. In the current chapter, we discuss protocols involved in immunofluorescence staining and identification of differentiated cells from BTIC populations.

Key words Differentiation assay, Fetal bovine serum, Immunofluorescence, Antibodies

1 Introduction

The differentiation assay is one of many in the armamentarium used to determine the presence of stem cell properties in the utilized cell line. Cell differentiation is a developmental process by which a less specialized cell becomes more specialized as it commits to a more specific cell lineage. In the case of cancer stem cells, these multipotent cancer stem cells would have the capacity to differentiate into all the cells of the tissue of tumor origin. In the example of brain tumor initiating cells (BTICs) from a primary brain tumor glioblastoma, we would expect to identify neurons, astrocytes, and oligodendrocytes [1, 2]. Each differentiated cell possesses cell-specific markers (such as GFAP for astrocytes, MAP2 for neurons, and Olig2 for oligodendrocytes) after differentiation that can be visualized under the microscope as well as fluorescence activated cell sorting (FACS) analysis. The advantage of microscopic analysis is the ability to simultaneously visualize the morphology in correlation to the cell-specific antibody marker staining.

Sheila K. Singh and Chitra Venugopal (eds.), *Brain Tumor Stem Cells: Methods and Protocols*, Methods in Molecular Biology, vol. 1869, https://doi.org/10.1007/978-1-4939-8805-1_8, © Springer Science+Business Media, LLC, part of Springer Nature 2019

2 Materials

Prepare all solutions using analytical grade reagents in a tissue culture biosafety cabinet. Prepare and store at 4°C (unless otherwise indicated). Follow all waste disposal regulations of your local institution and as per the reagent's MSDS.

1. IBIDI® plates or borosilicate cover slips.
2. BTIC cultures.
3. Paraformaldehyde (PFA, 4%).
4. Triton X-100.
5. Phosphate-buffered saline (PBS).
6. Primary antibodies
    (a) Neurons: MAP2/Tuj1.
    (b) Astrocytes: GFAP.
    (c) Oligodendrocytes: Olig2.
7. 488/546/647 secondary antibody.
8. Fetal bovine blocking serum—FBS (usually of the animal the secondary antibody is raised in).
9. Light protective container.
10. DAPI containing mounting media.
11. Poly-L-Ornithine Solution.
12. Laminin Solution.
13. Flow cytometer (MoFlo XDP, Beckman Coulter) with Flow cytometer analysis software (Kaluza, Beckman Coulter).
14. Automated cell counter or hemocytometer.
15. 2 mM EDTA in 1× DPBS.
16. 5 mL round-bottom polystyrene test tube with 35 μm cell strainer.
17. Fixable viability dye (LIVE/DEAD™ Fixable Near-IR Dead Cell Stain Kit).
18. BD/Cytofix/Cytoperm Kit.
19. Conjugated anti-human GFAP antibody.

3 Methods

Complete all procedures at room temperature unless otherwise specified.

3.1 Cell Differentiation for Microscopic Analysis

1. Prepare either IBIDI® plates or borosilicate cover slips for adherent culture propagation with poly-L-orinithine and laminin.

(a) Place 1× 10 mm coverslip into a 12-well plate. Prepare as many as required (*see* **Note 1**).

(b) Dilute poly-L-ornithine in a 1:4 dilutions into molecular grade distilled water and add enough to 12 well to coat the coverslip (i.e., 500 μL).

(c) Cover with lid and incubate plate in 37°C cell culture incubator for 1 h.

(d) Prepare required volume of diluted laminin solution. Using 1 mg/mL laminin concentration, aliquot 20 μL per 4 mL of PBS.

(e) Remove poly-L-ornithine solution and add diluted laminin solution into wells with coverslips.

(f) Cover with lid and incubate the plate in 37°C cell culture incubator for 2 h-overnight.

(g) If plates/cover slips are not going to be in use the following day, replace diluted laminin solutions with PBS and store in fridge (lid sealed with paraffin).

2. BTICs cultured in serum-free media are plated at 10% confluency. Recommend approximately 10×10^5 cells onto the appropriate sized borosilicate cover slip or IBIDI® plate.

3. Add fetal bovine serum at a concentration of 10% to your cell propagation media.

4. The cells are then allowed to differentiate for 2 weeks ensuring coverslips or IBIDI® wells are always covered by serum-containing media.

3.2 Microscopic Analysis

3.2.1 Coated Coverslips

1. Utilize borosilicate glass coverslips for best resolution. Note the thickness and the refractive index of the coverslips if utilizing the confocal microscope for analysis.

2. Slip coverslips into 12 or 24-well plate. Ensure bottom surface is flat.

3. Prepare coverslips or IBIDI® as described in 3.1 .

3.2.2 Fixing Cells

1. Once cells have been differentiated for >2 weeks, wash wells with PBS.

2. Add 4% PFA to wells ensuring cells are completely immersed.

3. Incubate at room temperature for 20 min.

4. Cells ready for permeabilizing (*see* **Note 2**).

3.2.3 Permeabilizing and Blocking Cells

1. Dilute Triton-X 100 to 0.05% and add to wells and incubate for 10 min at room temperature (*see* **Note 3**).

2. Wash cells with PBS × 3.

3. Add blocking solution (1.5% serum in PBS) to wells and incubate for 1 h in room temperature (*see* **Note 4**).

3.2.4 Primary Antibody Staining

1. The primary antibody of interest will usually have been validated by the company. For best results identify peer-reviewed references utilizing the antibody and dilute to the concentration used in the reference (*see* **Note 5**).

2. Prepare enough primary antibody in the blocking solution to cover the cells. Remember to keep wells for secondary only to be able to detect secondary background.

3. Add primary antibody solution to the appropriate wells. Place onto the rocker at low speed and incubate for 1 h in room temperature.

3.2.5 Secondary Antibody Staining

1. Wash the wells with PBS × 4 at 15-min intervals for 1 h.

2. After third wash, prepare secondary antibody in blocking solution (*see* **Note 6**).

3. Protect from light and add to wells. Ensure you have one well each of secondary only.

4. Incubate at room temperature for 45 min on a low speed rocker.

5. Following incubation, wash with PBS for 45 min at an interval of every 5 min while protecting from light.

6. If using IBIDI® slides can now take to confocal microscope to analyze. If using coverslips, drop 1× DAPI mounting media onto glass slides and place coverslips (cells down) onto the slide. Incubate at room temperature protected from light overnight. Slides are ready for microscopic analysis thereafter.

3.3 Flow Cytometric Analysis

3.3.1 Cell Preparation for FACS Analysis

1. Prepare cells using enzymatic dissociation such that they are in a single cell suspension.

2. Wash cells with 1× DPBS and spin at 290 × g for 5 min.

3. Resuspend the pellet with 2 mM PBS-EDTA and filter cell suspension into a polystyrene test tube with a 35 μm cell strainer.

4. Determine cell count and viability using an automated cell counter or a hemocytometer.

3.3.2 Intracellular Staining for GFAP Antibody

1. Incubate cell suspension with a select viability dye, which can be used for fixed cells, in the dark at 4 °C for 20 min.

2. Wash cells with 2 mM PBS-EDTA by centrifuging at 290 × g for 3 min, add 200 μL of BD Cytofix/Cytoperm Buffer™ (BD Cytofix/Cytoperm Kit) to the pellets, and incubate at 4°C for 20 min.

3. Wash cells with 1× BD Perm/Wash™ buffer by centrifuging at 290 × g for 3 min and resuspend in 200 μL of 1× BD Perm/Wash™ buffer.

4. Aliquot cells into two tubes, each containing 100 μL of cell suspension, and add fluorophore-conjugated anti-human GFAP antibody (e.g., GFAP-PE) at 1:10 dilution to one tube and matched isotype control to the other tube (*see* **Note 5**).

5. Incubate stained cell suspension in the dark at 4°C for 30 min.

6. Wash cells with 1× BD Perm/Wash™ buffer by centrifuging at $290 \times g$ for 3 min.

7. Resuspend pellets in approximately 250 μL of 2 mM PBS-EDTA, and analyze on flow cytometer.

3.3.3 Analysis via Flow Cytometry

1. Run an unstained sample to set a plot with forward and side scatter of cells on a flow cytometer using proper flow cytometry analysis software.

2. Set a proper gate for viable cells using a positive control for select viability dye.

3. Use single stained cell suspension or compensation beads to compensate for spectral overlap between fluorescence of viability dye and antibody conjugate(s).

4. Run cells stained with a matched isotype control to account for background fluorescence (Fig. 1).

5. Run stained samples and use gates set with controls to determine positive populations.

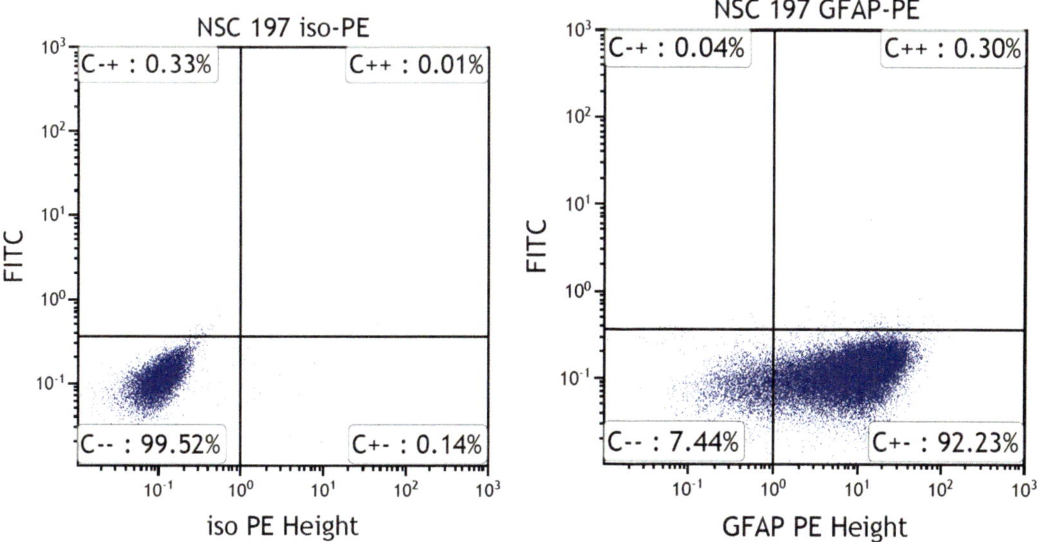

Fig. 1 Flow cytometric analysis of GFAP in neural stem cells differentiated in presence of fetal bovine serum. An isotype control antibody is used as a negative control to determine nonspecific binding (left). Human anti-GFAP antibody conjugated to phycoerythrin (PE) is used for analysis of protein expression in cells

4   Notes

1. Ensure preparation of enough wells to have all the antibodies probed and the secondary only probing in order to determine the appearance of the background from a secondary only (e.g., if probing for GFAP, Tuj1 and Olig2 all in one coverslip, you would need at least four coated coverslip with adherent cells).

2. If not permeabilizing cells immediately, immerse coverslip/ IBIDI® in PBS and store in fridge. It can be stored for up to 2 weeks for optimal results.

3. Permeabilization should not be completed if probing for surface proteins.

4. Blocking is completed with serum of the animal that the secondary antibody is raised in (i.e., utilize Goat serum if antibodies are raised in goat).

5. Dilution of the antibody will require optimization in the cells being utilized.

6. Take care to err on the side of lower concentrations to avoid nonspecific binding of secondary (Fig. 2).

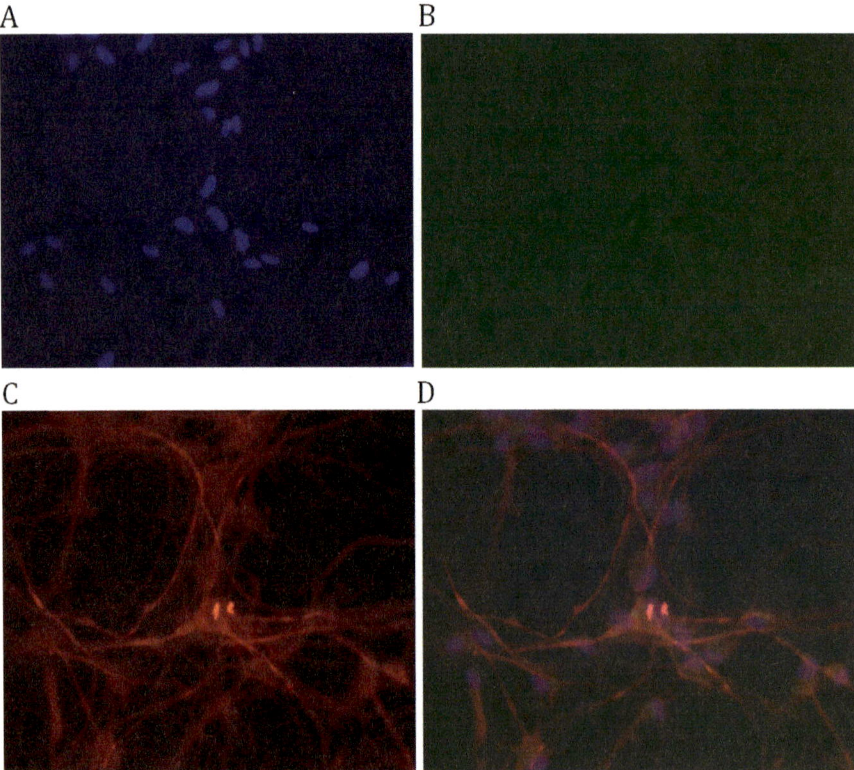

Fig. 2 Cerebellar neural stem cells stained with (**A**) nuclear staining DAPI, (**B**) cerebellar granule cellular marker Math1, (**C**) neuronal marker Tuj1 (D): Merge of all 3 antibody stainings

References

1. Singh SK et al (2004) Identification of human brain tumor initiating cells. Nature 432:396–401. https://doi.org/10.1038/nature03128

2. Singh S, Clarke I, Terasaki M, Bonn V (2003) Identification of a cancer stem cell in human brain tumors. Cancer Res 63(18):5821–5828

Chapter 9

The Study of Brain Tumor Stem Cell Migration

Montserrat Lara-Velazquez, Rawan Al-kharboosh, Luis Prieto, Paula Schiapparelli, and Alfredo Quiñones-Hinojosa

Abstract

Despite many advancements in brain cancer therapeutics, brain cancer remains one of the most elusive diseases with high migratory capacity and a dismal prognosis. It is well established that tumor stem cells utilize the same available migratory machinery that normal cells employ. Some of the major determinants of brain tumor stem cell *migration* are their cytoskeletal rearrangements and adhesion dynamics. This phenomenon allows brain tumor stem cells to perpetually migrate, invade, and repopulate in a vicious cycle leading to tumor expansion and invasion at tumor boundaries. In order to dissect the enabling factors that allow for this process to be hijacked, we have identified relevant assays to enable measurements of neoplastic migration such as Boyden Chamber, 3D chemogradient chamber, Nanopattern, and wound healing assays. Our purpose is to report the complex experimental platforms seen in the literature today and provide an optimal platform to kick off your studies in this field.

Key words BTICs migration, Transwell, Boyden chamber, Chemogradient chamber, Nanopattern, Wound healing assay

1 Introduction

Despite vigorous treatment combinations of surgery, chemotherapy, and radiation, there is almost a universal recurrence rate due to a small subset of cells known as brain tumor initiating cells (BTICs). BTICs can move as singular cells or move in clusters to infiltrate delicate matrices in various ways [1]. Their invasive capacity is enabled through the decisive manipulation of their shape, size, stiffness, and volume as well as expressing the pro-migratory signals that facilitate their motility. In order to delineate the precise factors responsible for your cells' motile capacity, we have highlighted some laboratory assays that have been widely used in the field. *Migration* is defined as the movement of cells through a substrate or a porous membrane in a predictable fashion with speed,

Montserrat Lara-Velazquez and Rawan Al-kharboosh contributed equally to this work

Sheila K. Singh and Chitra Venugopal (eds.), *Brain Tumor Stem Cells: Methods and Protocols*, Methods in Molecular Biology, vol. 1869, https://doi.org/10.1007/978-1-4939-8805-1_9, © Springer Science+Business Media, LLC, part of Springer Nature 2019

directionality, and persistence. Cancer initiating cell migration has been a significant interest in the field of brain tumor biology and substantial efforts have been put forth to identify the distinct molecular players underlying this hi-jacked process leading to eventual cancer metastasis [2]. BTICs can migrate "singularly" or "collectively" in a cohort establishing organized motile tumor masses that invade secondary organs [1, 3]. Most solid tumors use "collective migration" defined by a single cell via cadherins and other adhesion receptors to bind to the direct adjacent cell, while cell-cell communication manifests *via* gap-junctions [4]. Glioblastoma, a grade IV glioma for instance, travels through what is termed *"mesenchymal migration"* that is dependent on integrin associated adhesion dynamics and frictional forces navigating the leading and lagging edges mediating directional migration [4–6]. We have laid out a foundation of experimental assays that will resolve your curiosity of the mechanical capability of brain tumor initiating cell migration collectively and down to the single cell; it is essential to understand the engines involved in propelling cancer cells forward and how to stifle these very engines that took so many lives.

2 Materials

2.1 Boyden Chamber

1. Standard 24-well plate.

2. Transwell unit: predetermine pore size—8.0 μm polycarbonate membrane—6.5 mm TC-treated with lid, sterile.

3. BTIC Base media: 500 mL DMEM/F-12 1% antibiotic (5 mL antibiotic) add Neuroplex.

4. 10 mL Neuroplex without vitamin A serum-free supplement

5. Complete BTIC media: 50 mL basal media with 20 ng/mL each of EGF and FGF.

6. Q-tips.

7. Cotton swabs.

8. PBS without calcium and magnesium.

9. 4% paraformaldehyde.

10. Triton.

11. PBS with calcium and magnesium.

12. Fetal bovine serum.

13. VECTASHEILD mounting medium with DAPI.

14. Sterile pipette tips.

15. Sterile forceps.

16. Cell of interest to be seeded on top.

17. Charged white glass microscope slides.

18. Microscope cover slips.

19. AccuSharp disposable ophthalmic knives.

20. Chemoattractant: could consist of conditioned media, drug, serum, adherent cells at the bottom etc.

21. Trypsin.

22. 15 mL conical tubes.

23. Cell counter.

24. Nail polish.

2.2 3D Chemogradient Chamber

1. μ-Slide Chemotaxis, ibiTreat: #1.5 polymer coverslip, tissue culture treated, sterilized.

2. Matrigel suitable for cell culture.

3. Cells of interest.

4. Chemoattractant of interest.

5. A humid chamber to decrease evaporation (a petri dish with wet tissue can be used).

6. Inverted microscope with a $5\times$ objective and $10\times$ phase contrast.

7. Camera for time lapse movies and software for video processing.

8. Motorized stage for parallel data acquisition with auto focus.

9. CO_2 Incubation system.

10. Graduated pipettes from 10–200 μL.

11. Beveled pipette tips from 10–200 μL (Greiner bio-one).

2.3 Nanopattern

1. Multi-well nanopattern with parallel ridges 350 nm wide, 500 nm high, and spaced 1.5 μm apart.

2. 70% ethanol.

3. 100% ethanol.

4. Poly-D-lysine (10 μg/mL).

5. Mouse laminin 1 $\mu L/cm^2$.

6. Basement membrane (BBM) component.

7. Cell media.

8. Inverted microscope.

9. Temperature and gas controlled microscope chamber.

10. MATLAB script or your preferred software to allow manual tracking and measurement of cells frame by frame.

11. Phosphate-buffered saline (PBS) $1\times$.

12. Alconox.

13. Contact solution.

14. Windex.

15. Petri dish.

2.4 Wound Healing

1. Vi Cell counter machine.

2. 12-well plates culture dish.

3. Native or transfected cells.

4. DMEM X12

5. Accutase.

6. Phosphate-buffered saline (PBS) $1\times$.

7. Bovine serum albumin (BSA).

8. Mouse laminin.

9. Sterile 200 µL micropipette tip.

10. Serum-free DMEM.

11. Stage incubator.

12. CO_2 supply.

13. Digital camera connected to the inverted microscope

14. A software to analyze the captured image of your choice.

3 Methods

3.1 Boyden Chamber (Fig. 1)

1. Pre-warm media of interest that will be tested. This media will include both your native media of your tested cells as well as the chemoattractant media.

2. Harvest migratory cell of interest by washing cells to be studied (BTICs) from flask with PBS and shake flask back and forth to remove cell debris.

3. Aspirate PBS from flask.

4. Add trypsin or cell dissociation reagent and incubate for 5–10 min or until cells detach and floating in flask.

5. Collected floated cells.

6. Place collected cell in 15 mL conical tube.

7. Spin cells at $200 \times g$ from 5 min.

8. Aspirate the supernatant and resuspend pellet in 1 mL.

9. Count cell pellet to acquire final cell concentration in 1 mL.

10. Prepare desired concentration of chemoattractant +2% serum at 500–600 µL/well.

11. Assemble transwell inserts in your 24-well plate. Insert should be hanging above your 500–600 µL chemoattractant.

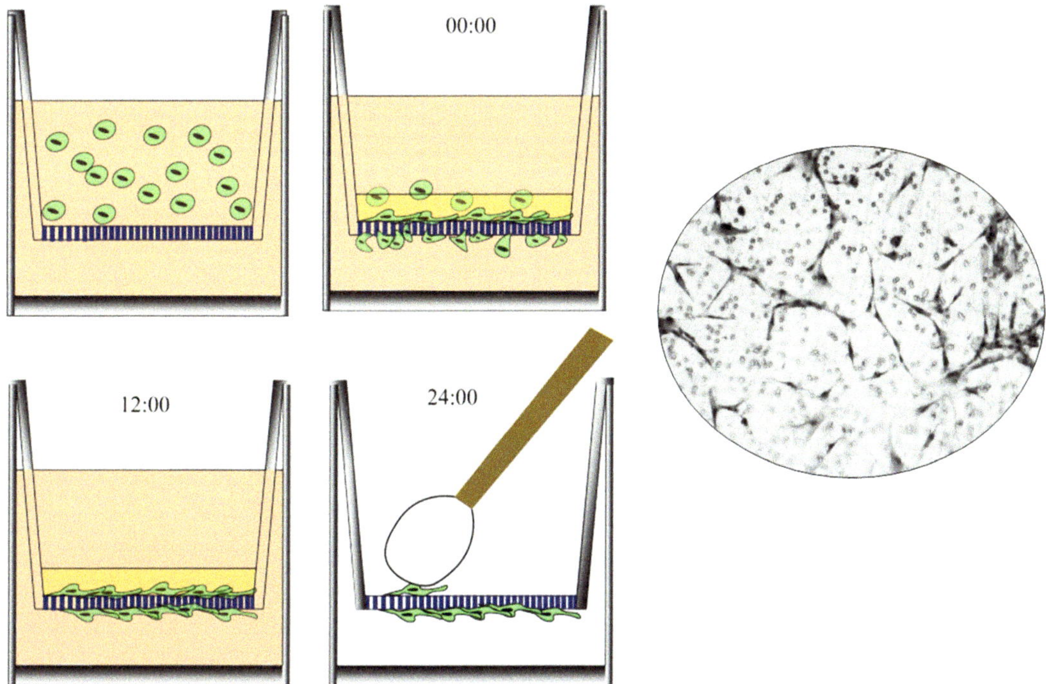

Fig. 1 Boyden Chamber: The cells are placed in a well divide by a membrane with pores of different sizes. The upper chamber is filled with the treated or conditioned media with cells and the bottom chamber with a chemo gradient component. The cells in the upper chamber will migrate vertically to cross the membrane toward the bottom chamber. Adapted with permission from: Brief Report: Robo1 Regulates the Migration of Human Subventricular Zone Neural Progenitor Cells During Development. Guerrero-Cazares H, Lavell E, Chen L, Schiapparelli P, Lara-Velazquez M, Capilla-Gonzalez V, Clements AC, Drummond G, Noiman L, Thaler K, Burke A, Quiñones-Hinojosa A. Stem Cells. 2017 Jul;35(7):1860–1865. doi: 10.1002/stem.2628. Epub 2017 Apr 24

12. Add 2×10^4 migratory cells (BTICs) inside transwell insert + 0.5% serum (FBS or FCS) for a final volume of 200 µL per insert.

13. Incubate in 37 °C and leave overnight for migration (recommended 22–24 h).

14. When cells are ready to be counted, remove the transwell insert from each well and follow the steps as follows, sequentially:

Migratory cell counting for boyden chamber:

1. Use a new 24-well plate.

2. Each row add.

    (a) Row 1: 500 µL PBS across all wells.

    (b) Row 2: 500 µL 4% PFA across all wells.

    (c) Row 3: 500 µL PBS/0.1% Triton across all wells.

    (d) Row 4: 500 µL PBS 1×.

3. Take out your 24-well plate with the migratory cells from incubator and remove each transwell.

4. Flip transwell over to remove all cells that have not migrated from the top (the inside of the transwell—at this point, all cells that have migrated are on the other side of the membrane facing the chemoattractant). All other cells that have not migrated are on the top of the membrane and inside the transwell where you first seeded them and need to be removed with a cotton swab.

5. Place a sterile cotton swab inside the transwell and gently scrape the membrane to remove all un-migrated cells making sure to get all corners.

6. Place transwell in row 1 to wash in PBS.

7. Remove transwell from the well in row 1 and clean inside with the cotton swab to remove any residual cells.

8. Place transwell in row 2 in 4% PFA for 5 min to fix cells.

9. Remove transwell from row 2, swab the top of the membrane with cotton swab again to remove any cells.

10. Place transwell in row 3 dipped in PBS/0.1% triton to permeabilize the cells for DAPI staining for quantification later in the protocol.

11. Remove transwell from row 3 and wipe inside to remove any residual non-migratory cells.

12. Place transwell in row 4 well in last step of PBS wash and swab.

13. Remove the transwell from row 4 and swab inside to remove any residual cells.

14. Take the AccuSharp disposable ophthalmic knives and cut the membrane making sure to visualize the side where the migratory cells are found (the bottom of the membrane).

15. Place the cut membrane with migrated cell side facing up, on a charged white glass microscope slides.

16. Place a drop of VECTASHEILD mounting medium with DAPI on the top of the membrane and allow drop to spread across the membrane making sure not to allow bubbles.

17. Place a microscope cover-slip on the top of the membrane slowly.

18. Seal coverslip on microscope slide using nail polish around the edges.

19. Visualize and count cells under an inverted microscope against controls.

3.2 3D Chemogradient Chamber (Fig. 2)

Unpack the μ-Slide Chemotaxis and put it into a humid chamber (a petri dish with a wet tissue inside)

1. Prepare cell suspension.

2. Resuspend a cell concentration of 3×10^6 cells in 200 μL of culture media.

3. Mix the cell suspension with 100 μL of Matrigel® (*see* **Note 1**).

4. Close filling ports C, D, E, and F with the plugs. Handle the plugs with tweezers.

5. Use a 20 μL precooled pipet and apply 6 μL of the suspension to the top of filling port A, leaving space between the tip and the port (*see* **Note 2**).

6. With a 6 μL pipet aspirate air from the opposite filling port B. Press the pipet tip directly into the port until the suspension reaches the pipet tip.

7. Leave both filling ports A and B filled with gel.

8. Gently remove all plugs from filling ports C, D, E, and F. Close filling ports A and B with plugs.

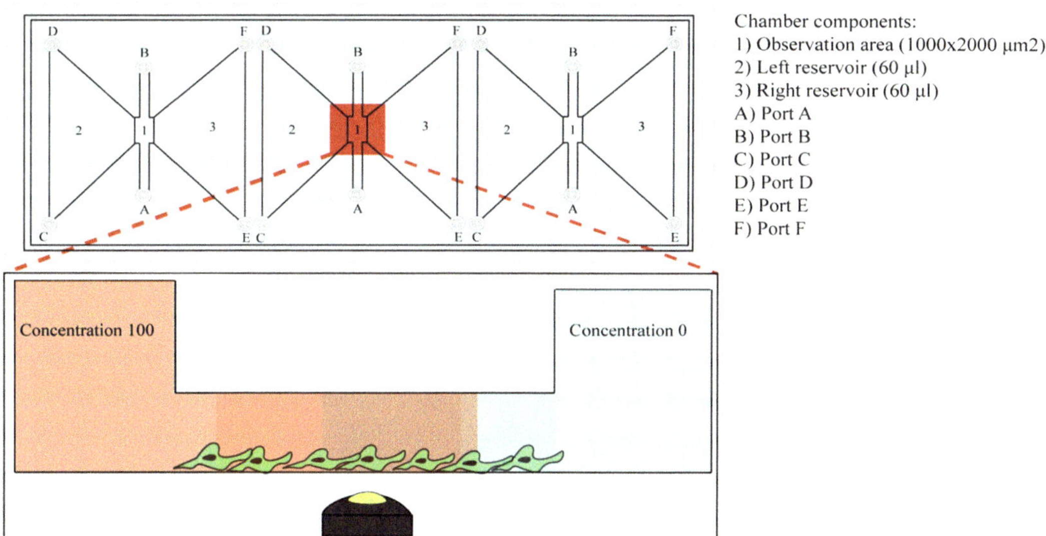

Fig. 2 3D Chemogradient Assay: Chemogradient chamber enables the study of cellular migratory responses to chemical cues. The platform is subdivided into chambers where the observational chamber is coated to allow for cell attachment. A chemoattractant or a chemorepellent is put into one chamber and allowed to diffuse toward the other chamber on the opposite end. The chamber separation causes a chemo-gradient in a linear fashion whereby the chemo-factor concentration progressively decreases as you move further away from the chamber where the chemo-factor is added. This assay allows for the identification of chemotactic or chemorepellent activity of your agent of interest and the parallel assessment of volumetric contribution of your soluble factor to brain tumor migration in a dose-dependent manner

9. Incubate at 37 °C the slide inside the petri dish to minimize evaporation until the gel is formed (approximately 30 min).

10. Control the cell morphology with a microscope during and after gelation.

11. Gently close filling ports C and D with plugs (chemoattractant side).

12. Fill the first reservoir by injecting 65 μL of the chemoattractant-free medium through filling port E. Keep in mind that filling ports E and F must be completely filled, but not overfilled.

13. Transfer the two plugs from the filling ports C and D to the filling ports E and F. This will close the chemoattractant-free side.

14. Fill the empty reservoir by injecting 65 μL of chemoattractant-free medium through filling port C.

15. Now the chamber is completely filled with chemoattractant-free medium and cells will only grow inside the gel in the observation area. Control your cells under the phase contrast microscope.

16. Use a 20 μL pipet (e.g., Gilson P-20) and apply 15 μL chemoattractant to the top of filling port C. Do not inject directly.

17. Aspirate 15 μL liquid from the opposite filling port D. Press the pipet tip directly into filling port D. The chemoattractant on the top of filling port C will be flushed inside and fill the reservoir.

18. Repeat the two mentioned steps again to load 30 μL of chemoattractant.

19. Gently close all filling ports.

Time-lapse

20. After selected incubation period, place the slide in an inverted microscope stage.

21. Set up a time-lapse experiment according to the selected conditions.

22. Export your images for analysis.

3.3 Nanopattern (Fig. 3)

Cleaning nanopattern before experiment

1. Use sterile fluids for all steps.

2. Wash nanopattern wells with 500 μL PBS.

3. Pipette up and down to rinse pattern.

4. Aspirate PBS from the corners taking care not to scratch the well.

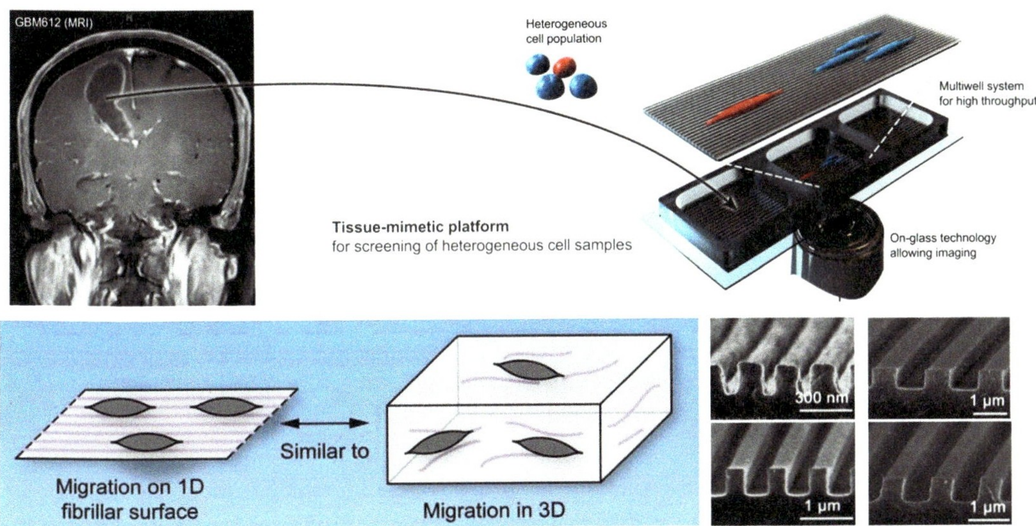

Fig. 3 Nanopattern assay. Cancer cells with heterogeneous phenotypes are isolated from a patient's tumor and placed in a smooth or patterned surface for a period of time. Imaging of migration and morphology with single-cell resolution can be capture to study the migratory behavior of the cancer cells . Adapted with permission from: Migration Phenotype of Brain-Cancer Cells Predicts Patient Outcomes. Chris L.Smith, OnurKilic, PaulaSchiapparelli2, HugoGuerrero-Cazares, Deok-HoKim, Neda I.Sedora-Roman, SakshamGupta, ThomasO'Donnell, Kaisorn L.Chaichana, Fausto J.Rodriguez, SaraAbbadi, JinSeokPark, AlfredoQuiñones-Hinojosa, AndreLevchenko. Cell Reports. 15(12),21. June 2016.Pages: 2616-2624

5. Add 500 µL alconox, pipette up and down and incubate for 15–30 min.

6. Wash with 500 µL sterile water, pipette up and down then aspirate.

7. Add 500 µL of sterile contact solution, pipette up and down then aspirate.

8. Incubate at room temperature overnight.

9. Next day, aspirate sterile contact solution and wash with water, then aspirate.

10. Wash with 500 µL Windex. Pipette up and down and aspirate.

11. Wash with 500 µL of 70% ethanol, pipette up and down and aspirate.

12. Let air dry for a few minutes.

13. Store in petri dish until later use.

Nanopattern preparation

1. Coat 8-well nanopattern plate with poly-D-lysine and incubate at room temperature for 15 min (use 400 µL per well).

2. Aspirate and coat with laminin solution at 400 µL/well.

3. Incubate laminin at room temperature for 1 h.

Fig. 4 Wound Healing Assay: The wound healing assay allows the study of cellular migration through a scrape in a cellular culture monolayer. The cells on the edges of the scrape will migrate to close the wound, re-establishing cell to cell connections. (Modified from: The agar diffusion scratch assay - A novel method to assess the bioactive and cytotoxic potential of new materials and compounds. Mascha Pusnik, Minire Imeri, Grégoire Deppierraz, Arie Bruinink, and Manfred Zinn. Sci Rep. 2016; 6: 20854)

4. Aspirate laminin and put 400 μL of PBS per well until ready.

5. Seed cells to be tested at $4–5 \times 10^4$/well in a 400 μL volume.

Nanopattern imaging software

1. Choose imaging software and preferred time-lapse parameters.

2. Time points suggested.

   (a) Time point: 60.

   (b) Duration: 10 h.

   (c) Interval: 10 min.

3. Count 100 cells per condition.

3.4 Wound Healing (Fig. 4)

1. Coat the cell culture dishes with laminin/PBS $1\times$ dilution (20 μg/mL) to cover all the work surface, incubate the dishes at $+37\ °C$, 5% CO_2 for 2 h if a quick coating is needed or overnight for a slow coating (*see* **Note 3**).

2. Aspirate the laminin with a glass pipette, without disturbing the coated surface and refill the dishes with 3–5 mL of pre-warmed media before seeding the cells. If there is interest in a specific compound, this should be added into the media before seeding of cells as well (*see* **Note 4**).

3. Seed the required number of cells into the prepared dish.

4. Place the dish in incubation at $+37\ °C$, 5% CO_2. Depending on the nature of the cell line, 1 or 2 days would be needed to have a confluent monolayer.

5. With a 200 pipet tip scrap the center of the cell monolayer in a straight line to create a "scratch" (*see* **Note 5**).

6. Gently wash the cells once with 1 mL of the medium to remove the debris, then, replace with 3–5 mL of medium specific for the in vitro wound healing assay.

7. Determine the time period to monitor the scratch area (usually 0–48 h) (*see* **Note 6**).

8. Place the dish in a tissue culture incubator at 37 °C under a phase-contrast microscope and set the monitor session determining the reference points and the time frame in the microscope software (*see* **Note 7**).

9. Take the first picture and align the photographed region acquired in **step 6** and acquire a second image.

10. Monitor migration distance of cells in each well using Meta-Morph software, version: 6.1. (Carl Zeiss, Germany).

11. Analyze and compare the images from time 0 to the last time point.

4   Notes

1. To decrease Matrigel density, tweezers, pipet tips, and the matrigel itself need to be maintained cold (on ice).

2. Do not inject the suspension (matrigel and cells) with the pipet placed directly into the port to avoid bubble formation.

3. The coated plates need to be placed in a flat surface without shaking, to ensure a uniform coating of the surface.

4. Culture medium solutions should be pre-warmed in a bath tank at +37 °C, 5% CO_2 before use.

5. The width and height of the scratches in the monolayers cultures need to be approximately of similar size between groups (the assessed cells and control cells) to minimize variation.

6. To obtain the same field during the image acquisition, set the reference points close to the scratch. After selection of the reference points, place the dish under a phase-contrast microscope and acquire the first image of the scratch.

7. To select a time frame of incubation, migratory capabilities of the cells and the input of the tested conditions must be considered to achieve a complete closure of the scratch.

   It is recommended to measure at least 100 cells.

Acknowledgments

AQH was supported by the Mayo Clinic Professorship and a Clinician Investigator award as well as the NIH (R43CA221490, R01CA200399, R01CA183827, R01CA195503, R01CA216855). MLV was supported by CONACYT and PECEM from the National Autonomous University of Mexico. We would like to thank Hugo Guerrero-Cazares for his contribution to figures in this chapter.

References

1. Smith CL, Kilic O, Schiapparelli P, Guerrero-Cazares H, Kim DH, Sedora-Roman NI et al (2016) Migration phenotype of brain-Cancer cells predicts patient outcomes. Cell Rep 15 (12):2616–2624

2. van Zijl F, Krupitza G, Mikulits W (2011) Initial steps of metastasis: cell invasion and endothelial transmigration. Mutat Res 728(1–2):23–34

3. Friedl P, Wolf K (2010) Plasticity of cell migration: a multiscale tuning model. J Cell Biol 188 (1):11–19

4. Petrie RJ, Yamada KM (2012) At the leading edge of three-dimensional cell migration. J Cell Sci 125(Pt 24):5917–5926

5. Nina Kramera AW, Ungera C, Rosnera M, Krupitzab G, Hengstschlägera M, Dolznig H (2012) In vitro cell migration and invasion assays. Mutat Res 752:10–24 ElSevier

6. Tilghman J, Schiapparelli P, Lal B, Ying M, Quinones-Hinojosa A, Xia S et al (2016) Regulation of Glioblastoma tumor-propagating cells by the integrin partner Tetraspanin CD151. Neoplasia 18(3):185–198

Chapter 10

The Study of Brain Tumor Stem Cell Invasion

Rawan Al-kharboosh, Montserrat Lara-Velazquez, Luis Prieto, Rachel Sarabia-Estrada, and Alfredo Quiñones-Hinojosa

Abstract

Ninety percent of deaths from solid tumors have been ascribed to the invasion and metastatic dissemination of cancer cells. One of the most fundamental prerequisites of brain tumor stem cell *invasion* is their ability to penetrate and traverse through the basement membrane and stromal compartments displacing from their original point of origin and repopulating at a distant site and adjacent tissue. In order to propose a successful clinical strategy, the investigation of the molecular determinants of invasion must factor in measurements and predictors of the complexity of brain tumor invasion potential accounting for the physiological scenarios encountered in the tumor niche. This chapter will highlight some laboratory approaches such as: spot assay, 3D chemogradient chambers, Fluoroblok™ tumor invasion, organotypics, and 3D tumor spheroid assays that could help mitigate the investigation of a tumor stem cell's invasion capacity through restrictive 3D environments otherwise seen in vivo.

Key words Motility, BTICs, Invasive capacity

1 Introduction

There is growing evidence that a small subset of cells, termed brain tumor initiating cells (BTICs), are responsible for new tumor formation due to their enhanced invasive and migratory capacity, enabling them to undergo a multistep process by augmenting natural cellular mechanisms that allow them to traverse across tissues in a perpetual and sometimes unstoppable manner. *Invasion* is defined as the destruction of surrounding matrices in a 3D system (in vivo) or 3D matrix (in vitro) requiring breakdown or lysis of extracellular components, manipulation of adhesion molecules, and remodeling of obstructive fiber networks that contain X, Y, and Z planes a BTIC must navigate through [1, 2]. This multi-step process requires distinct genetic players permitting (1) detachment of BTICs from the primary cancer, (2) destructing surrounding fibers

Rawan Al-kharboosh and Montserrat Lara-Velazquez contributed equally to this work.

Sheila K. Singh and Chitra Venugopal (eds.), *Brain Tumor Stem Cells: Methods and Protocols*, Methods in Molecular Biology, vol. 1869, https://doi.org/10.1007/978-1-4939-8805-1_10, © Springer Science+Business Media, LLC, part of Springer Nature 2019

and matrices, (3) intravasation through vessels entering the circulatory system or lymphatic system, (4) extravasation from circulation, (5) settlement, proliferation, and eventual outgrowth of new tumor at secondary site [3].This step-wise process requires cytoskeleton mediated deviations starting with enhanced capacity to (1) polarize the body in a directional manner with the leading edge at the front and the lagging at the back, (2) this dynamic change potentiates a cell protrusion morphology resulting in a pseudopod, lamellipod, filopod, invadopods, and a few other phenotypes enabling contractile forwarding or retraction of the cell body, (3) vigorous interaction with actin and components of the extracellular matrix, (4) adherence and lysing of structural component, and finally (5) detachment of the trailing edge leading to propulsive movement forward [1, 4, 5]. This chapter will identify experimental assays for researchers in the field who desire to investigate migration and invasion in cancer; understanding such masked responses to environmental cues will help illuminate the causes giving rise to cancer invasiveness owing to the dismal prognosis seen in cancer patients today.

2   Materials

2.1   Spot Assay

1. Cells to be tested (must be labeled).
2. Eppendorf tubes.
3. Horse serum.
4. DNase I.
5. 6-well plate.
6. Polycarbonate filters (with pore size of choice, we recommend size smaller than your cell of interest 1–3 μm).
7. PBS.
8. 4% PFA.
9. HBSS.
10. Pipettes.
11. Cell dissociation media (trypsin/accutase).
12. Culture media.
13. Neurobasal media (NB).
14. L-glutamine.
15. Penicillin/streptomycin.
16. B27 supplement.

2.2 Fluoroblok^{TM} Tumor Invasion for Coculture

1. Cells to be tested (cell type A).
2. Trypsin/Accutase/cell dissociation reagent.
3. Media of migratory cells (without serum).
4. Cells releasing Chemoattractant/soluble factors (cell type B).
5. DPBS.
6. Corning Fluoroblok™ cell culture inserts (24-well 8.0 µm pore size).
7. 24-well Fluoroblok™ support.
8. Fetal bovine serum.
9. Calcien AM fluorescent.
10. DMSO.
11. HBSS ++.
12. Fluorescence plate reader with bottom reading capabilities.
13. Matrigel matrix.
14. Coating buffer for matrigel: 0.01 M Tris (pH 8.0), 0.7% NaCl.
15. 0.2 µm sterile filter unit.
16. 15 mL conical tube.
17. Ice.
18. Sterile forceps.
19. Sterile pipette tips.

2.3 Organotypics

1. Laminar flow culture hood.
2. Dissecting microscope.
3. HBSS plus Ca and Mg.
4. Microsurgical instruments (forceps, scissors, and scalpel).
5. Autoclaved Pasteur pipettes.
6. 10 mm petri dishes.
7. 24-well culture plate.
8. Organotypic culture media: Minimal Essential Medium (MEM) (sigma, St. Louis, MO, USA) containing 25% heat-inactivated horse serum (Gibco/BRL, Bethesda, MD), 25% HBSS (Gibco/BRL, Bethesda, MD) with 25.8 mg/mL of glucose, and 12 mg/mL of 4-(2-hydroxyethyl)-1-piperazineethanesulfonic acid (HEPES) buffer (sigma, St. Louis, MO, USA) and 1% 0.2 M, glutamine (Gibco/BRL, Bethesda, MD) pH 7.2.
9. Microtome or vibratome capable of 300 µm tissue thickness.
10. 12 mm culture plate inserts (Millipore, Billerica, MA).
11. Angled dissecting microscope or surgical microscope.

2.4 3D Tumor Spheroid Invasion	1. Cell line tested (must form spheres which BTICs usually do).

2.4 3D Tumor Spheroid Invasion

1. Cell line tested (must form spheres which BTICs usually do).

2. Ultra-low attachment 96-well plate.

3. 5× basement membrane extract (BME) or matrigel coating solution, keep at 4 °C.

4. Culture medium of cell to be tested.

5. Cytometer or any form of cell counter.

6. Inverted microscope.

7. Sterile pipettes.

8. PBS.

9. Cell dissociation solution.

10. Wash buffer.

11. Laminar flow and incubator.

12. Multichannel pipette.

13. Ice bucket.

3 Methods

3.1 Spot Assay (Fig. 1)

3.1.1 Making of Neurobasal Media (NB)

1. DMEM/F12 HEPES.

2. 2 mM L-glutamine.

3. B27 serum-free supplement.

4. 1% penicillin/streptomycin.

5. hEGF 20 ng/mL.

6. hFGF 10 ng/mL.

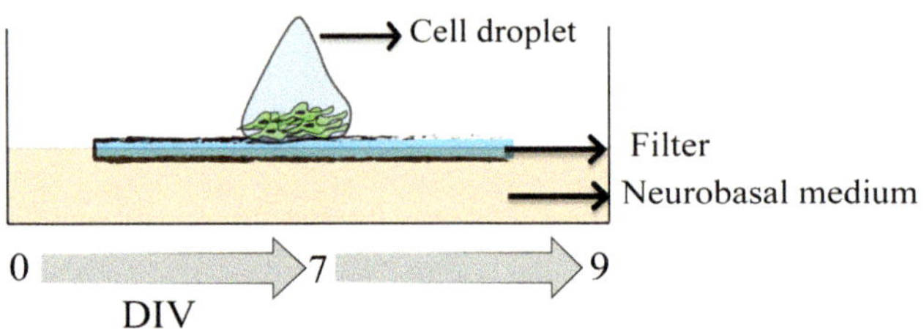

Fig. 1 Spot Assay: The assessment of migration is dependent on the distance traveled from the middle spot of which the cells were dropped. The principal relies on placing a polycarbonate filter in the middle of a well and dropping 1ul of cell concentration on the top of the filter. Over time, the cell of interest will migrate away from the point of origin whereby the filter is then fixed, imaged and analyzed for distance cells traveled in response to exogenous supplements. Adapted from: Dose-dependent effect of EGF on migration and differentiation of adult subventricular zone astrocytes. Perez-Gonzalez O, Quinones-Hinojosa A. Glia.58 (8),2010.Pages:975–983

3.1.2 Preparation of Cell Transplantation

1. Dissociate labeled cells to be tested from flask.
2. Spin down dissociated cells according to pelleting protocols.
3. Wash cell pellet in NB medium supplemented with 10 μg/mL DNase I.
4. Centrifuge 5 min, $200 \times g$.
5. Remove media.
6. Resuspend pellet with NB for a final concentration of 1×10^5 cells per μl.

3.1.3 Spot Assay Drop

1. Place 1.5–2 mL of NB medium inside each well of a 6-well plate.
2. Place one polycarbonate filter inside each well allowing it to float.
3. Place 1 μL drop of cell suspension in the middle of the filter.
4. Decide your time point (note: allow for a number of days in vitro (DIV) to assess the migratory capacity of cells: recommended 24–72 h).
5. When cell migration time point is reached, replace media with 4% PFA fixative at 37 °C for 1 h.
6. Take the plate, incubate it at 4 °C for 6 h in the same fixative.
7. Image your cells in a fluorescence microscope.
8. Photograph each replicate.
9. Use imaging software (ex: NIH IMAGE) for cell migration analysis.
10. Assess distance of furthest cell traveled from center point.

3.2 Fluoroblok™ Tumor Invasion for Coculture (Fig. 2)

1. Prepare coating buffer: 0.01 M Tris (pH 8.0), 0.7% NacL. Filter using a 0.2 μm sterile filter unit.
2. Thaw matrigel on ice.

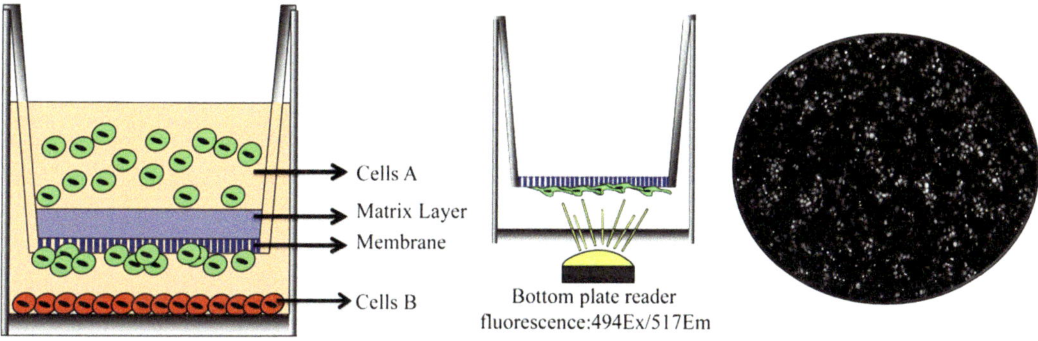

Fig. 2 Fluoroblok invasion assay. This technique follows the same premise as the transwell migration assay except the porous membrane is overlaid with a layer of a matrix a cell must transfer through

3. Once matrigel is thawed, swirl it several times to ensure it is evenly dispersed.

4. Prepare coating solution to place inside transwell by mixing matrigel at a final concentration of 200–300 μg/mL with coating buffer until the final volume comes to 2.0 mL and place solution on ice until ready for use.

5. Prepare calcein AM labeling by diluting stock solution with DMSO for a final concentration of 1 mg/mL.

6. Prepare working solutions of calcein for a final volume of 4 μg/mL and use 500 mL/well in HBSS++ for a 2 μg/well concentration for cell quantification.

3.2.1 Permeable Support Coating of Matrigel

1. Under sterile conditions, assemble transwells inside 24-well support plates to prepare for matrigel coating.

2. With ice cold pipette tips, carefully add 0.1 mL of the diluted matrigel coating to each permeable membrane. Avoid air bubbles and minimize contact with walls.

3. Incubate plates with coated matrigel inside transwells at 37 °C for 2 h. Do not let matrigel matrix layer dry out.

4. Prepare an equal number of controls.

3.2.2 Preparation of Cell A (Cell to Be Tested) and Cell Type B (Cell to Be Invaded)

1. Prepare cell suspension in culture medium for cell A and cell B.

2. Spin and pellet cells at 1×10^5 cells/mL for cell A.

3. Determine optimal dilution/concentration for cell type B *(A: B --1:2 or 1:3. Meaning if you plate cell type A at 1×10^5 cells/mL, for a 1:2 concentration, your cell B seeded at the surface of well would be 2×10^5 cells/mL).

4. Add the 1:2 or 1:3 dilution of cell B at the bottom for a final cell volume of 500 μL/well.

5. Add 200 μL/well of cell A (cells invading the matrix) inside well and on the top of a matrigel invasion chamber.

6. Make sure there are no air bubbles.

7. Incubate invasion chambers overnight (22–24 h) in a humidified tissue culture at 37 °C, 5% CO2 atmosphere (*see* **Note 1**).

3.2.3 Measurement of Cell Invasion

1. When ready to measure invasion, carefully remove medium from apical membrane chambers which can be accomplished by careful aspiration.

2. It is advisable to use forceps to flip and flick the inside of the contents inside a biohazard bin. Do not touch the bottom surface of the insert system.

3. After removing the inside contents of the invasion chambers, transfer the inserts in another 24-well support fluoroblok plate containing 500 μL of 4 μg/mL Calcein AM in HBSS++.

4. Incubate for 1 h at 37 °C, 5% CO_2 (*see* **Note 2**).

5. For a bottom plate reader, it is recommended that the dimensions of your plate and the gains setting must be adjusted to quantitate fluorescent average appropriately.

6. Start with a midpoint setting and allow the bottom plate reader to read fluorescence calcien AM moiety.

7. Fluorescence of invaded cells is read at wavelength 494 excitation and 517 emission on a bottom-reading fluorescent plate reader.

8. Data is expressed as % invasion = mean RFU of cells invaded through matrigel matrix/mean RFU of cells migrated through uncoated membrane (*see* **Note 3**).

9. The results could be verified by using an inverted fluorescent microscope and is especially helpful to do this the first time you do this experiment.

3.3 **Organotypics** *(Fig. 3)*

1. Tissue sample (human or rodent) must be transported in saline solution, PBS, or cell culture media, and kept on ice until processing.

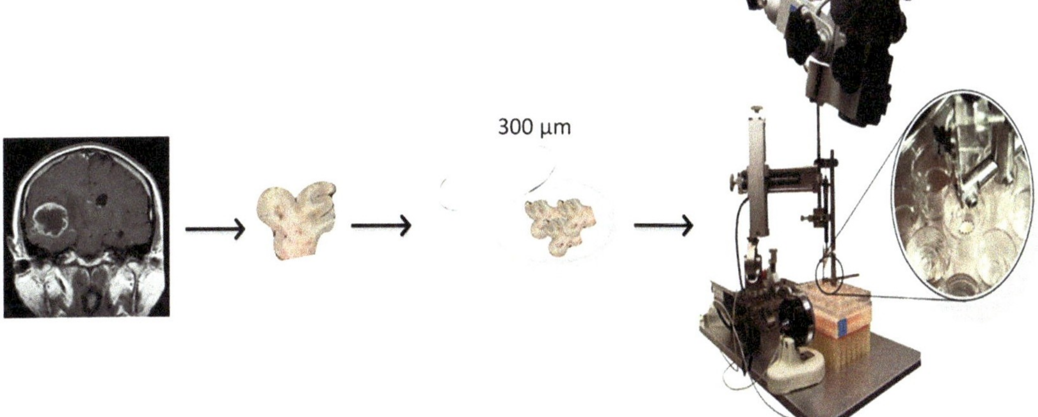

Fig. 3 Human Organotypics. The organotypic assay is a powerful in vitro system that allows for the specific analysis of brain cell migration and invasion ex vivo. Organotypic tissue slices are 300um in thickness and stay viable for several days. One could study the mechanical capabilities of cells in a platform that maintains brain cytoarchitecture in a 3D fashion. Brain tumor cells are labeled and incorporated on top of the tissue slice where their penetration and displacement over several days are assessed and measured in a 3 dimensional manner using a two-photon microscope. Adapted from: Neurosphere culture and human organotypic model to evaluate brain tumor stem cells. Guerrero-Cázares H, Chaichana KL, Quiñones-Hinojosa A. Methods Mol Biol. 2009;568:73-83. doi: 10.1007/978-1-59745-280-9_6. Preservation of glial cytoarchitecture from ex vivo human tumor and non-tumor cerebral cortical explants: A human model to study neurological diseases. Chaichana KL, Capilla-Gonzalez V, Gonzalez-Perez O, Pradilla G, Han J, Olivi A, Brem H, Garcia-Verdugo JM, Quiñones-Hinojosa A. J Neurosci Methods. 2007 Aug

2. Inside of a culture hood, place the dissecting microscope, the sterile surgical instruments, and the 10 mm petri dish (all the instruments need to be previously cleaned with ethanol).

3. Fill a petri dish half full with the HBSS high glucose solution under the hood.

4. Place the tissue sample into the filled petri dish and with the surgical instruments remove the necrotic areas and blood vessels from the tissue under the dissecting microscope.

5. Remove organotypic culture media from the refrigerator and place under the hood for 20–30 min to pre-warm the solution.

6. Using sterile technique in the hood, add 1 mL pre-warm organotypic culture media into each 24-well culture plate, and using forceps place a 12 mm Millipore insert into each incubation media-filled well. Cover the plate and set aside in the hood (*see* **Note 4**).

7. With the tissue chopper, cut the tissue to have slices of thickness of 350 microns.

8. Place the slides into a new petri dish containing high-glucose HBSS.

9. Carefully separate each transverse section of tissue with curved tweezers or by squirting solution onto the tissue and transfer each individual section of tissue onto the Millipore inserts.

10. Place approximately a tissue section in each well.

11. The plate with the organotypic cultures is then placed into an incubator at 37 °C and 5% CO_2.

12. The organotypic culture media is carefully changed every 2 days (*see* **Note 5**).

3.4   3D Tumor Spheroid Invasion (Fig. 4)

1. Pre-warm media along with chemoattractant media (conditioned media/cytokine media/media with your soluble factor of interest).

2. Harvest migratory cell of interest by washing cells to be studied (BTICs) from flask with PBS and shake flask back and forth to remove cell debris.

3. Aspirate PBS from flask.

4. Add trypsin or cell dissociation reagent and incubate for 5–10 min or until cells detach and floating in flask.

5. Collect floated cells.

6. Place collected cells in a 15 mL conical tube.

7. Spin cells at $200 \times g$ from 5 min.

8. Aspirate the supernatant and resuspend pellet in 1 mL.

9. Count cell pellet to acquire final cell concentration in 1 mL.

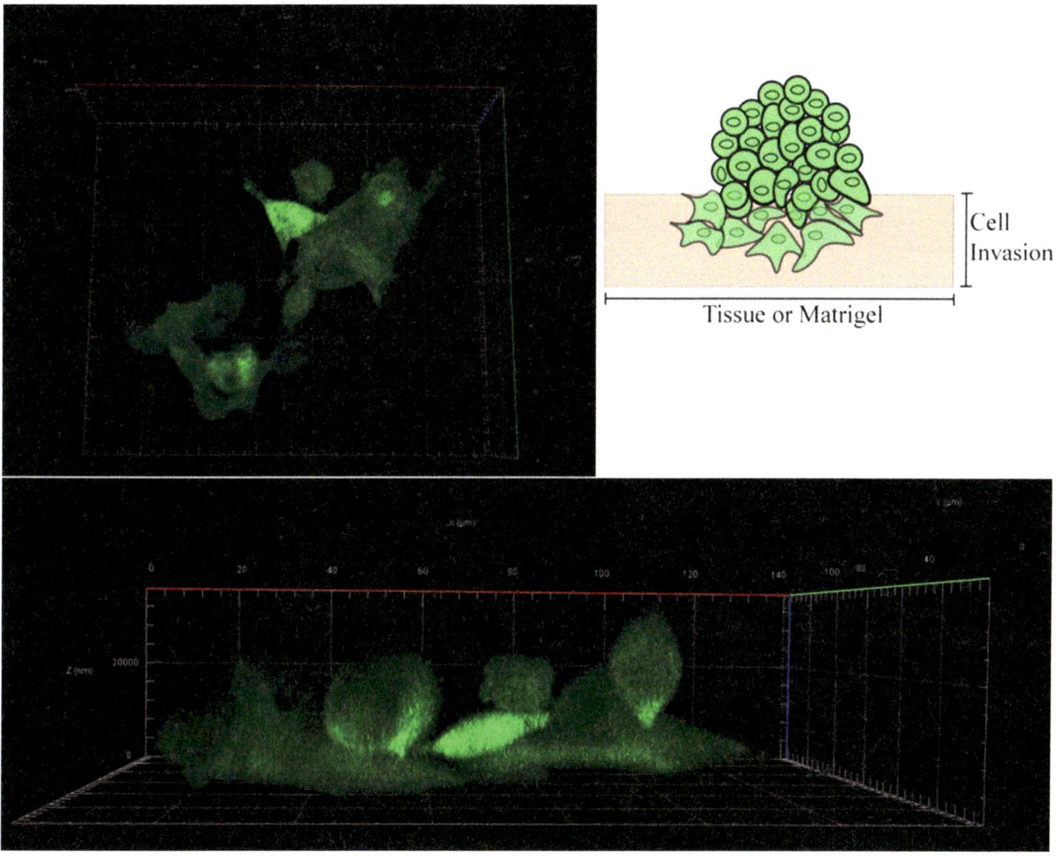

Fig. 4 3D Tumor Spheroid Invasion: The principal of this assay is simple and heavily dependent on brain tumor stem cell propensity to aggregate in spheres. The tumor spheroid is embedded in a basement like setting usually composed of extracellular matrix components and entrapped within a well. The invasive capacity of these cells are measured by assessing the ability of a single cell to break from the tumor sphere bulk and invade adjacent space by evaluating the distance traveled from the point of origin. The analysis is based on sequential time point images of the tumor spheroid taken over a period of 3-4 days and analyzed through imaging software to determine distance traveled. Figures donated from Hugo Guerrero-Cazares

10. For tumor spheroid per well, obtain $0.5–2 \times 10^4$ cells/mL and optimize seeding so that the diameter of spheroid formation after 4 days is seen to be around 300–500 μm.

11. Transfer the cell suspension to a sterile reservoir and, using a multichannel pipette, dispense 200 μL/well into ultra-low attachment (ULA) 96-well round-bottom plates.

12. Transfer the plates to an incubator (37 °C, 5% CO_2, 95% humidity).

13. Visually confirm tumor spheroid formation after 3–4 days and proceed with the 3D invasion assay.

3.4.1 Preparation of BMM

1. Thaw BMM on ice.

2. Allow sterile filter tips to be on ice overnight.

3. Place the 4-day old spheres from the 96-well plates on ice.

4. Gently remove 100 µL old media using a multichannel pipette. Make sure to angle the tips in such a way so as to not disrupt the spheroid and/or create bubbles.

5. Using the cold pipette tips, transfer BMM to ice cold tubes (*see* **Note 6**).

6. Gently dispense 100 µL of BMM into the U-bottom 96-well plate aiming the tip toward the inside of the well to avoid spheroid disturbance.

7. Repeat all steps for all groups and allow for at least $n = 5$ replicates.

8. To remove any bubble formations, use a sterile needle to remove or pop them.

9. Using a microscope, visually check the position of the spheroid as it must be positioned in the middle of the well (*see* **Note 7**).

10. After visualizing spheroid and confirming central position, transfer the plate to incubator and allow BMM extract to solidify.

11. After 1 h, gently add 100 µL of growth medium. At this point, you can add cytokines/drugs/growth medium/etc. for screening and testing reagents.

3.4.2 Image Acquisition and Analysis

1. Use an inverted microscope at $10\times$ objective (or $4\times$ if invaded area is too large).

2. Record spheroid image per well at intervals starting from $t = 0$ up to $t = 96$ h.

3. Save each image acquired, individually.

4. Open stage graticule image on your software analysis of choice.

5. Input measurements: units/objective.

6. For calibration, take images of a stage graticule (1 mm) using both $10\times$ and $4\times$ objectives (and save images).

7. Measure the area of spheroid covered by outlining the distance from $t = 0$ origin to your endpoint.

8. Export measurements of different parameters (area, diameter, perimeter, etc.) to a spreadsheet. Record on the spreadsheet the relevant image information (e.g., well number, relative time point).

9. Plot the mean area of replicate spheroids (invasion) versus time using scientific graphing and statistical software. Alternatively, calculate the change in spheroid area at each time point, relative to the area at $t = 0$. Then plot invasion (% t_0) versus time as a linear graph.

4   Notes

1. Depending on your A: B dilution, you may want to optimize parameters to determine if migratory and invasive capacity is enough to augment an overnight migration. It is also acceptable to include a serum gradient to potentiate migration by adding 2–5% FBS/FCS.

2. The Calcein AM solution is not removed from the lower chamber before reading fluorescence because it is a non-fluorescent vital dye that is converted into green fluorescent calcein by cytosolic esterases—although it is recommended.

3. The Fluoroblok chambers effectively block the passage of light from 490–700 nm at >99% efficiently, labeled cells that have not invaded do not have to be removed such as the earlier Boyden chamber described in this chapter. The cells that have invaded through the matrix and are found at the bottom of the membrane are no longer shielded from the light source and are thus detected.

4. Slowly place the inserts into each well to avoid bubble formation.

5. The media needs to be changed using autoclave Pasteur pipettes attached to a vacuum system. It is recommended refilling each insert with 1 mL of organotypic media.

6. It is recommended for your tested cell line to use growth factor to stimulate invasion as you would use a serum gradient to stimulate migration.

7. If spheroid is not found in middle, then centrifuge the plate at $300 \times g$ for 3–5 min at 4°°C to allow for central positioning of spheroids.

Acknowledgments

AQH was supported by the Mayo Clinic Professorship and a Clinician Investigator award as well as the NIH (R43CA221490, R01CA200399, R01CA183827, R01CA195503, R01CA216855). MLV was supported by CONACYT and PECEM from the National Autonomous University of Mexico. We would like to thank Hugo Guerrero-Cazares for his contribution to figures in this chapter.

References

1. Nina Kramera AW, Ungera C, Rosnera M, Krupitzab G, Hengstschlägera M, Dolznig H (2012) In vitro cell migration and invasion assays. Mutat Res 752:10–24 ElSevier

2. Smith CL, Kilic O, Schiapparelli P, Guerrero-Cazares H, Kim DH, Sedora-Roman NI et al (2016) Migration phenotype of brain-Cancer cells predicts patient outcomes. Cell Rep 15 (12):2616–2624

3. van Zijl F, Krupitza G, Mikulits W (2011) Initial steps of metastasis: cell invasion and endothelial transmigration. Mutat Res 728(1–2):23–34

4. Smith CL, Chaichana KL, Lee YM, Lin B, Stanko KM, O'Donnell T et al (2015) Pre-exposure of human adipose mesenchymal stem cells to soluble factors enhances their homing to brain cancer. Stem Cells Transl Med 4 (3):239–251

5. Abbadi S, Rodarte JJ, Abutaleb A, Lavell E, Smith CL, Ruff W et al (2014) Glucose-6-phosphatase is a key metabolic regulator of glioblastoma invasion. Mol Cancer Res 12 (11):1547–1559

Chapter 11

Cell Cycle Dynamics in Glioma Cancer Stem Cells

Ingrid Qemo and Lisa A. Porter

Abstract

Cancer stem cells, sometimes referred to as tumor initiating cells, play pivotal roles in tumor initiation, progression, metastasis, resistance to therapy, and relapse. Understanding how these populations of cells expand in response to a host of conditions is critical in determining effective cancer therapeutics. A defining feature of cancer stem cells is the ability to switch between modes of quiescence and symmetric/asymmetric division to protect and conserve the population, this feature is traditionally reserved for normal adult stem cell populations. Understanding how the core cell cycle machinery responds to external cues to drive symmetric/asymmetric division vs. quiescence will reveal fundamental information about how cancer stem cell populations survive and expand. This chapter will describe methods to study the cell cycle dynamics in brain cancer stem cell populations and how they compare to the other populations in a tumor.

Key words Glioma, Cancer, Stem cells, CD133, CD15, CD44, Cell cycle, Flow cytometry

1 Introduction

Cancer stem cells (CSCs) share many defining characteristics that are traditionally associated with normal stem cells. Normal multipotent adult stem cells exhibit key features, including high self-renewal capacity and the ability to differentiate into the specialized cells in the tissue where they reside [1]. Historically, these cells are thought to divide very infrequently in vivo to preserve their populations; however, the rate of division is highly dependent on the tissue type and external cues from the external environment (developmental or stress/injury related) [2–4]. When cued to replenish, adult stem cells can divide either symmetrically or asymmetrically [5, 6]. Mathematical modeling supports that symmetric division may occur in response to accumulating DNA damage to protect the adult stem cell population and that CSCs favor symmetric over asymmetric division to facilitate tumor growth [7–10]. Ultimately these decisions depend on the core cell cycle machinery, how CSCs alter the cell cycle to control this balance is an active area of research.

Sheila K. Singh and Chitra Venugopal (eds.), *Brain Tumor Stem Cells: Methods and Protocols*, Methods in Molecular Biology, vol. 1869, https://doi.org/10.1007/978-1-4939-8805-1_11, © Springer Science+Business Media, LLC, part of Springer Nature 2019

The cell cycle is a master regulatory network that is involved in the growth and division of cells, organismal development, and certain diseases, like cancer. The cell cycle can be divided into four distinct stages referred to as G1 (gap 1), S (synthesis), G2 (gap 2), and M (mitosis). The progression from one cell cycle phase to another is regulated by many proteins, most notably, the cyclin dependent kinases (CDKs) and their cyclin partners [11]. Cells may also exit the cell cycle and reside in the G0 phase. Cells enter this phase during adverse conditions, or in order to help preserve integrity and function over a prolonged period of time. The G0 phase can be entered from the G1 phase, where information is processed from the external environment as well as internal ques. [12]. A G0 phase in which a cell may reenter the cell cycle is referred to as a quiescent state. Adult tissue stem cells, including those residing in the brain, are found in a quiescent state until they are activated in response to cues, such as tissue damage [13]. The balance that occurs between G1 and G0 phases is responsible for the maintenance of a healthy stem cell pool while allowing lineage commitment to occur when necessary. The failure to maintain this balance is linked to tumorigenesis [5]. In cancer, the implications of a cell population that has the ability to indefinitely exit the cell cycle into a state of quiescence are alarming, and include the ability to survive grim environmental conditions and resist cytotoxic drugs [14]. However, it is important to note that the cell cycle dynamics of these cell populations in vivo most likely won't be reflective of their dynamics in vitro, due to the missing external environment.

The brain contains adult tissue stem cells that are referred to as neural stem cells (NSCs) which possess many similarities to the CSC population within Glioblastoma (GBM) the most common form of brain tumor, referred to as Brain tumor stem cells (BTSCs) [15–18]. GBMs contain a population of quiescent cells which contribute to their aggressiveness [19]; however, the cell cycle dynamics of this population remain relatively poorly characterized. Normal NSCs and GBM CSC populations express characteristic cell surface proteins, which are useful in identifying and sorting these cell populations in order to study or enrich them [20]. One of these markers, CD133, isolates populations of GBM CSC that have the capacity to recapitulate a tumor, and the presence of these cells correlates with patients that have a poor clinical prognosis [21, 22].

Other markers have been used to study, isolate, and enrich GBM CSCs, including CD15 (also called SSEA-1 [stage-specific embryonic antigen 1]) [23], CD44 [24, 25], and the ATP-binding cassette (ABC) multidrug transporters which contribute to chemoresistance and drug efflux capability [26]. Whether any of these GBM CSC markers correlate to cellular quiescence in vivo, or proliferative capacity in vitro has not been conclusively determined and may reveal information about the aggressiveness of select CSC populations.

One of the most rudimentary approaches to analyzing cell cycle dynamics is to measure the cellular DNA content [27]. This approach discloses whether a cell is in G0/G1 (2n), S (2n~4n), or G2/M (4n). A number of fluorescent dyes are readily available to use in order to label DNA content. Common fluorescent DNA dyes include DAPI, Hoechst 33342, and Propidium Iodide (PI), among many others. Different techniques have their own pros and cons, and, depending on the experiment, one technique may be more advantageous than the others [28]. Hoechst 33342 binds to the minor groove of DNA, specifically to adenine-thymine-rich areas and will emit a fluorescence when bound to the dsDNA. Furthermore, compared to its counterparts listed above, Hoechst 33342 is membrane permeant and is able to stain live cells.

This chapter describes the protocols to uncover cell cycle patterns in GBM CSC populations using various identified CSC markers and the DNA dye Hoechst 33342. The use of CD133, CD15, and CD44 in conjunction with Hoechst 33342 will help reveal different glioma CSC populations in vitro, in both established GBM cell lines and primary patient-derived GBM tumor cell lines, and allow for comparative analysis of cell cycle profiles.

2 Materials

2.1 Culturing Human SJ-GBM2

1. The human SJ-GBM2 glioblastoma cancer cell line was originally obtained from the Children's oncology group repository (Texas Tech University Health Sciences Center).

2. Iscove's modified Dulbecco's medium (IMDM) which contains 4 mM L-glutamine.

3. Fetal bovine serum (FBS).

4. Insulin- Transferrin- Selenium (100×) (ITS).

5. 100 × 20 mm polystyrene cell culture dishes.

2.2 Culturing Human Primary Patient-Derived GBM Cells

1. Primary human primary patient-derived glioblastoma multiforme neurospheres (HBTC 3160) were generously donated from Dr. deCarvalho at the Hermelin brain tumor center at Henry Ford Hospital.

2. Dulbecco's modification of Eagle's medium (DMEM)/Ham's F12 50/50 mix.

3. N-2 supplement (100×).

4. Epidermal growth factor (EGF) recombinant human protein (*see* **Note 1**).

5. Fibroblast growth factor (FGF)—Basic human, recombinant expressed in *E. Coli* (*see* **Note 1**).

6. Gentamicin solution.

7. Antibiotic-Antimycotic ($100\times$).

8. Cell culture grade bovine serum albumin (BSA), heat shock fraction, protease free, low endotoxin.

9. T25 tissue culture flasks.

2.3 Flow Cytometry: Cell Preparation and Fixation

1. Dulbecco's phosphate-buffered saline (D-PBS, $1\times$) solution without calcium and magnesium, pH 7.4.

2. Paraformaldehyde powder (95%) was used to make a 2% formaldehyde (PFA) solution in $1\times$ PBS.

2.4 Flow Cytometry: Staining and Analysis

1. Allophycocyanin (APC) conjugated Anti-human CD133/AC133 (Prominin-1) monoclonal antibody (TMP4) (eBioscience; cat. no 17-1338-42) (*see* **Note 2**).

2. The Alexa Fluor 555 mouse anti-SSEA-1 (CD15) antibody (MC480) (BD biosciences; cat. no 560119) (*see* **Note 2**).

3. Phycoerythrin (PE) conjugated Anti-CD44 monoclonal antibody [F10-44-2] (Abcam; cat. no ab82529) (*see* **Note 2**).

4. Hoechst 33342, trihydrochloride, trihydrate—10 mg/mL solution in water (thermo fisher scientific; cat. no H3570) (*see* **Note 3**).

5. Pharmingen stain buffer.

6. Axygen 1.5 mL eppendorf microcentrifuge tubes.

7. EASYstrainer 70 μm sterile cell strainers.

8. Polystyrene round-bottom 12×75 mm BD falcon tubes appropriate for flow cytometric analysis.

9. The analysis of the cells was preformed using a BD LSRFortessa X-20 flow cytometer with a BD FACSDiva software (BD Biosciences).

3 Methods

3.1 SJ-GBM2 Cell Preparation

1. SJ-GBM2 are cultured in IMDM which contains 4 mM L-glutamine and supplemented with 20% FBS, and $1\times$ ITS on 100×20 mm polystyrene cell culture dishes at 37 °C and 5% CO_2.

2. When cells reach 70–80% confluency, remove the media from cell culture dish and discard it.

3. Add 3 mL of cold Ca^{2+}/Mg^{2+}—free D-PBS solution to each cell culture dish, which will aid in rounding and detaching the cells from the dish. Observe cells under a bright field microscope until the majority of the layer is detached (*see* **Note 4**). It should take approximately 5 min.

4. Gently wash the cell layer and add it to a 15 mL conical centrifuge tube. Spin down the cells at 500 × g for 7 min at 4 °C.

5. Remove the supernatant and resuspend the cell pellet in 1 mL Ca^{2+}/Mg^{2+}—free D-PBS.

6. Add a 1:1 (v:v) ratio of SJ-GBM2 cells to trypan blue solution.

7. Count the number of live cells using a hemocytometer (*see* **Note 5**).

8. Spin the cells and resuspend in cold stain buffer to get the desired concentrations depending on which of the surface antigens you will be using (*see* **Note 6**).

3.2 HBTC 3160 Cell Preparation

1. HBTC 3160 cells are cultured in DMEM/F-12 50/50 culture media supplemented with 1× N-2 supplement, 20 ng/mL EGF, 10 ng/mL FGF, 50 μg/mL gentamicin solution, 1× antibiotic-antimycotic, and 0.5 mg/mL BSA in T25 flasks at 37 °C and 5% CO_2.

2. Pipette all the contents of the flask to a 15 mL conical centrifuge tube. Spin down the cells at 800 × g for 5 min at 4 °C.

3. Remove the supernatant and resuspend the cells in 2 mL cold Ca^{2+}/Mg^{2+}—free D-PBS solution. Gently pipette to dissociate the spheres.

4. Incubate the cells in cold Ca^{2+}/Mg^{2+}—free D-PBS solution at room temperature for 10 min, while gently pipetting every 2 min.

5. Add a 1:1 (v:v) ratio of HBTC 3160 cells to trypan blue solution.

6. Count the number of live cells using a hemocytometer (*see* **Note 5**).

7. Spin the cells and resuspend in cold stain buffer to get the desired concentrations depending on the surface antigens to be tested (*see* **Note 6**).

3.3 Flow Cytometry Fixation and Staining

1. Label all falcon tubes, including respective controls and samples for each cell line (*see* Table 1).

2. For APC conjugated CD133, both the SJ-GBM2 and HBTC 3160 should be stained with 5 μL of antibody in a final volume of 100 μL which contains 500,000 cells (*see* **Note 7**). Incubate cells for 1 h at 4 °C while gently pipetting cells every 5–10 min to prevent aggregation.

3. For PE conjugated SSEA1 (CD15), both the SJ-GBM2 and HBTC 3160 should be stained with 5 μL of antibody in a final volume of 50 μL which contains 500,000 cells (*see* **Note 7**).

Table 1
Controls and samples of flow cytometry experiment: the table shows important controls and the sample tubes used in this costaining experiment which are needed for accurate data analysis for each cell line

Tube number	Name	Antibody
1	Unstained	No antibody
2	Single stain	CD133-APC
3	Single stain	SSEA1 (CD15) - Alexa Fluor 555
4	Single stain	CD44-PE
5	Single stain	Hoechst
6	Double stain	CD133-APC, Hoechst
7	Double stain	SSEA1-Alexa Fluor 555, Hoechst
8	Double stain	CD44-PE, Hoechst

Incubate cells for 30 min at 4 °C while gently pipetting cells every 5 min.

4. For PE conjugated CD44, both the SJ-GBM2 and HBTC 3160 should be stained with 5 μL of antibody in a final volume of 100 μL which contains 500,000 cells (*see* **Note 7**). Incubate cells for 30 min at 4 °C while gently pipetting cells every 5 min.

5. After staining for the cell surface antigen, wash the cells twice in 1 mL cold Ca^{2+}/Mg^{2+}—free D-PBS solution. Spin down at 1000 rpm (1600 × g) for 5 min at 4 °C.

6. After the second wash, remove the supernatant and resuspend the cells in 500 μL cold 2% PFA in 1× PBS for 10 min at room temperature to fix the cells (*see* **Notes 8–10**).

7. Add 1 mL of cold Ca^{2+}/Mg^{2+}—free D-PBS solution to each tube and spin down at 1000 rpm for 5 min at 4 °C.

8. Discard the supernatant.

9. For tubes that only contain a single stain, resuspend the cells in 500 μL cold Ca^{2+}/Mg^{2+}—free D-PBS solution and filter this through a 70 μm cell strainer before adding the filtered contents to falcon tubes.

10. For tubes that need to be counterstained for DNA content, resuspend the cells in 500 μL cold Ca^{2+}/Mg^{2+}—free D-PBS solution that contains Hoechst 33342 at a dilution of 1:2000. Incubate tubes for 10 min at room temperature (*see* **Note 11**).

11. Add 1 mL cold Ca^{2+}/Mg^{2+}—free D-PBS solution to wash and spin down cells at 1000 rpm for 5 min at 4 °C.

12. Discard the supernatant and resuspend the cells in 500 μL cold Ca^{2+}/Mg^{2+}—free D-PBS solution and filter this through a

70 μm cell strainer before adding the filtered contents to BD falcon tubes.

13. Keep all tubes on ice and protected from light until it is time to analyze them by flow cytometry.

3.4 *Flow Cytometry Analysis*

1. All flow cytometry experiments were performed using BD LSRFortessa X-20 with FACSDiva software. Cell debris was excluded from analysis.

2. The Hoechst 33342 dye is excited at 355 nm and its fluorescence emission is measured at 460–490 nm.

3. The APC fluorochrome is excited at 633–647 nm and its fluorescence emission is measured at 660 nm. A representative analysis of CD133 positive and negative populations in both cell lines showing their cell cycle profiles is presented in Fig. 1.

4. The Alexa Fluor 555 fluorochrome is excited at 555 nm and its fluorescence emission is measured at 565 nm. A representative analysis of SSEA1 (CD15) positive and negative populations in both cell lines showing their cell cycle profiles is presented in Fig. 2.

5. The PE fluorochrome is excited at 488 nm and its fluorescence emission is measured at 575 nm. A representative analysis of CD44 positive and negative populations in both cell lines showing their cell cycle profiles is presented in Fig. 3.

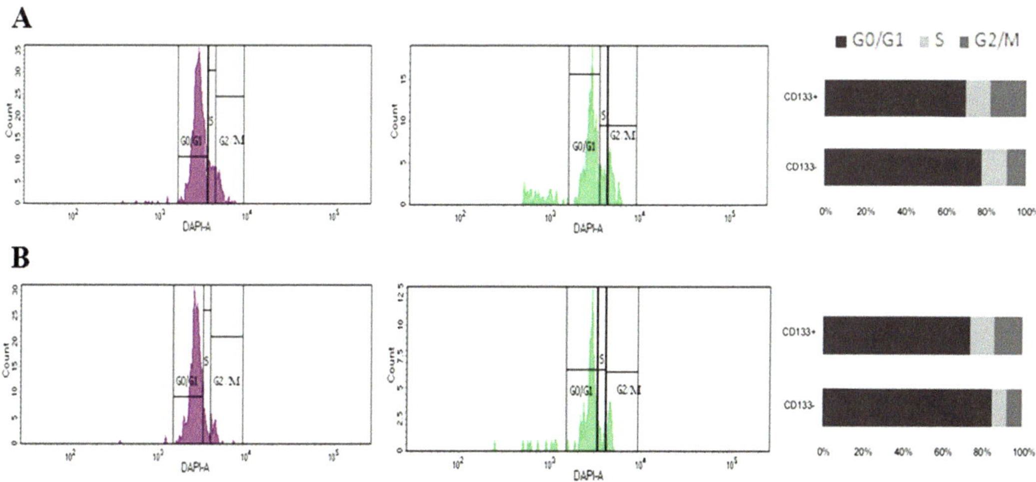

Fig. 1 Flow cytometry analysis of cell cycle profiles between CD133 negative (−) (purple) and CD133 positive (+) (green) populations in SJ-GBM2 glioblastoma cancer cells (**a**) and HBTC 3160 primary patient-derived glioblastoma cancer cells (**b**) and their respective graphs. Percentage of cells in GO/G1 decreases in the CD133+ population as compared to the CD33- population

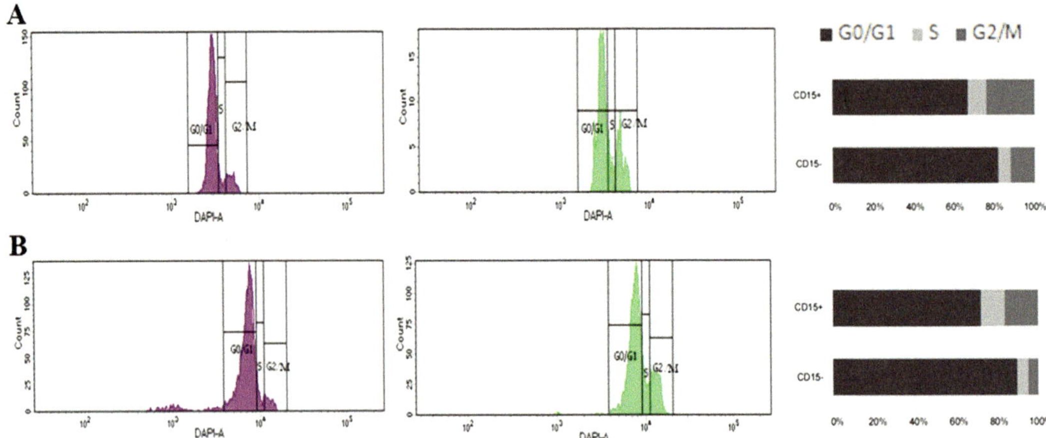

Fig. 2 Flow cytometry analysis of cell cycle profiles between CD15 negative (−) (purple) and CD15 positive (+) (green) populations in SJ-GBM2 glioblastoma cancer cells (**a**) and HBTC 3160 primary patient-derived glioblastoma cancer cells (**b**) and their respective graphs. Percentage of cells in G0/G1 decreases in the CD15+ population as compared to the CD15− population

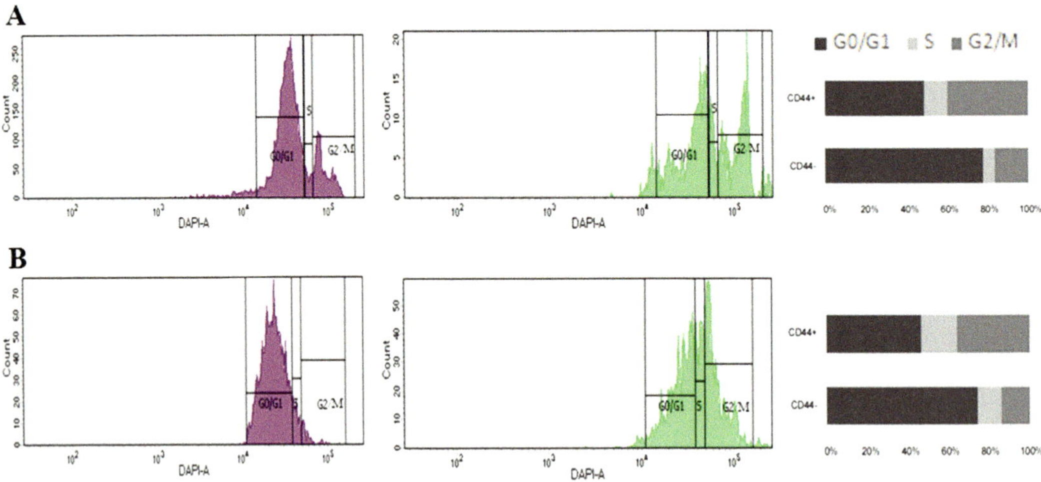

Fig. 3 Flow cytometry analysis of cell cycle profiles between CD44 negative (−) (purple) and CD44 positive (+) (green) populations in SJ-GBM2 glioblastoma cancer cells (**a**) and HBTC 3160 primary patient-derived glioblastoma cancer cells (**b**). Percentage of cells in G0/G1 decreases in the CD44+ population as compared to the CD44- population

4 Notes

1. In order to have no product loss, reconstituted EGF and FGF solutions should not be filtered. This is due to the fact that they may bind to the filters and there is no way to detect whether they have been rinsed from the filters.

2. Alexa Fluor 555 mouse anti-SSEA1 (CD15) antibody (MC480) reacts with both human and mouse samples according to the manufacturer's guidelines.

3. These antibodies come as solutions and should be protected from light.

4. If you find that the cells do not detach, you may alternatively consider using Accutase. However, care must be taken with the enzymatic solutions used, as this is a critical step for flow cytometry and could negatively affect the expression of surface antigens.

5. For optimal flow cytometry data and analysis you should aim for over 90% cell viability.

6. Cold solutions as well as staining at 4 °C is important in preventing the internalization or potential modulation of cell surface antigens.

7. All cell surface antigen staining should be performed in a 1.5 mL eppendorf tube and protected from light.

8. The order of staining and fixation varies depending on the vicinity of antigens: surface or intracellular antigens. When analyzing cell surface epitopes, it is best to stain these surface antigens first and then fix the cells afterward.

9. Be cautious on the fixation method that you choose. Fluorochromes like PE and APC are large and may be affected by alcohol fixation. We recommend formaldehyde fixation.

10. Fixing cells is an exothermic reaction and therefore you should aim to use cold fix solution.

11. Hoechst 33342 is membrane permeant, thus, a permeabilization step is not necessary.

References

1. Venere M, Fine HA, Dirks PB, Rich JN (2011) Cancer stem cells in gliomas: identifying and understanding the apex cell in cancer's hierarchy. Glia 59:1148–1154

2. Alvarez-Buylla A, Lim D (2004) For the long run: maintaining germinal niches in the adult brain. Neuron 41:683–686

3. Guilak F, Cohen D, Estes B, Gimble J, Liedtke W et al (2009) Control of stem cell fate by physical interactions with the extracellular matrix. Cell Stem Cell 5:17–26

4. Li L, Clevers H (2010) Coexistence of quiescent and active adult stem cells in mammals. Science 327:542–545

5. Morrison SJ, Kimble J (2006) Asymmetric and symmetric stem-cell divisions in development and cancer. Nature 441:1068–1074

6. Shen Q, Goderie SK, Jin L, Karanth N, Sun Y et al (2004) Endothelial cells stimulate self-renewal and expand neurogenesis of neural stem cells. Science 304:1338–1340

7. Shahriyari L, Komarova NL (2013) Symmetric vs. asymmetric stem cell divisions: an adaptation against cancer? PLoS One 8(10): e76195

8. Yatabe Y, Tavaré S, Shibata D (2001) Investigating stem cells in human colon by using 618 methylation patterns. Proc Natl Acad Sci 98:10839–10844

9. Spradling A, Drummond-Barbosa D, Kai T (2001) Stem cells find their niche. Nature 414:98–104

10. Nicolas P, Kim KM, Shibata D, Tavaré S (2007) The stem cell population of the

human colon crypt: analysis via methylation patterns. PLoS Comput Biol 3:e28

11. Norbury C, Nurse P (1992) Animal cell cycles and their control. Annu Rev Biochem 61:441–470

12. Zetterber A, Larsson O, Wiman KG (1995) What is the restriction point? Curr Opin Cell Biol 7(6):835–842

13. Cheung TH, Rando TA (2013) Molecular regulation of stem cell quiescence. Nat Rev Mol Cell Biol 14:329–340

14. Moore N, Lyle S (2011) Quiescent, slow-cycling stem cell populations in cancer: a review of the evidence and discussion of significance. J Oncol 2011:396076

15. Hemmati HD, Nakano I, Lazareff JA, Masterman-Smith M, Geschwind DH, Bronner-Fraser M, Kornblum HI (2003) Cancerous stem cells can arise from pediatric brain tumors. Proc Natl Acad Sci 100 (25):15178–15183

16. Ignatova TN, Kukekov VG, Laywell ED, Suslov ON, Vrionis FD, Steindler DA (2002) Human cortical glial tumors contain neural stem-like cells expressing astroglial and neuronal markers in vitro. Glia 39(3):193–206

17. Rosenblum ML, Gerosa M, Dougherty DV, Reese C, Barger GR, Davis RL, Levin VA, Wilson CB (1982) Age-related chemosensitivity of stem cells from human malignant brain tumours. Lancet 1(8277):885–887

18. Singh SK, Clarke ID, Terasaki M, Bonn VE, Hawkins C, Squire J, Dirks PB (2003) Identification of cancer stem cells in human brain tumors. Cancer Res 63(18):5821–5828

19. Campos B, Gal Z, Baader A, Schneider T, Sliwinski C, Gassel K (2014) Aberrant self-renewal and quiescence contribute to the aggressiveness of glioblastoma. J Pathol 234:23–33

20. Brescia P, Ortensi B, Fornasari L, Levi D, Broggi G, Pelicci G (2013) CD133 is essential for glioblastoma stem cell maintenance. Stem Cells 31:857–869

21. Zeppernick F, Ahmadi R, Campos B, Dictus C, Helmke BM, Becker N, Lichter P, Unterberg A, Radlwimmer B, Herold-Mende CC (2008) Stem cell marker CD133 affects clinical outcome in glioma patients. Clin Cancer Res 14:123–129

22. Han M, Guo L, Zhang Y, Huang B, Chen A, Chen W, Liu X, Sun S, Wang K, Liu A, Li X (2016) Clinicopathological and prognostic significance of CD133 in Glioma patients: a meta-analysis. Mol Neurobiol 53:720–727

23. Son MJ, Woolard K, Nam DH, Lee J, Fine HA (2009) SSEA-1 is an enrichment marker for tumor-initiating cells in human glioblastoma. Cell Stem Cell 4:440–452

24. Liu G, Yuan X, Zeng Z, Tunici P, Ng H, Abdulkadir IR, Lu L, Irvin D, Black KL, Yu JS (2006) Analysis of gene expression and chemoresistance of CD133$^+$ cancer stem cells in glioblastoma. Mol Cancer 5:67

25. Brown DV, Filiz G, Daniel PM, Hollande F, Dworkin S et al (2017) Expression of CD133 and CD44 in glioblastoma stem cells correlates with cell proliferation, phenotype stability and intra-tumor heterogeneity. PLoS One 12(2): e0172791

26. Bleau AM et al (2009) PTEN/PI3K/Akt pathway regulates the side population phenotype and ABCG2 activity in glioma tumor stem-like cells. Cell Stem Cell 4:226–235

27. Darzynkiewicz Z, Huang X (2004) Analysis of cellular DNA content by flow cytometry. Curr Protoc Immunol 5:7 Chapter 5: Unit

28. Kim KH, Sederstrom JM (2015) Assaying cell cycle status using flow cytometry. Curr Protoc Mol Biol 111:28.6.1–28.6.11

Chapter 12

Embryonic Stem Cell Models of Human Brain Tumors

Ludivine Coudière Morrison, Nazanin Tatari, and Tamra E. Werbowetski-Ogilvie

Abstract

Utilization of human embryonic stem cells (hESCs) as a model system to study highly malignant pediatric cancers has led to significant insight into the molecular mechanisms governing tumor progression and has revealed novel therapeutic targets for these devastating diseases. Here, we describe a method for generating heterogeneous populations of neural precursors from both normal and neoplastic hESCs and the subsequent injection of neoplastic human embryonic neural cells (hENs) into intracerebellar or intracranial xenograft models. Histopathologically, neural tumors derived from neoplastic hENs exhibit features similar to more aggressive medulloblastoma, the most common malignant primary pediatric brain tumor. In this chapter, we will outline the detailed methods for culturing normal and neoplastic neural precursor cells in both adherent and tumorsphere format and the full characterization of the brain tumors generated from these cells in non-obese diabetic severe combined immunodeficiency (NOD SCID) mice.

Key words Human embryonic stem cells, Neural precursors, Medulloblastoma, NOD SCID, Intracranial transplantation, Intracerebellar transplantation

1   Introduction

Human embryonic stem cells (hESCs) share several characteristics with cancer cells including upregulation of oncogene expression, increased proliferation, genomic instability, and elevated telomerase activity [1]. Following long-term adaptation in culture, hESCs also acquire major karyotypic changes that prohibit their use in the clinic [2]. We have previously shown that hESCs undergo neoplastic transformation in culture by adopting phenotypic changes such as growth factor independence, enhanced self-renewal, and aberrant differentiation both in vitro and in vivo [3]. These tumorigenic features are also sustained following generation of neural precursors (NPs) from the transformed hESCs, as the cells continue to exhibit enhanced proliferation and decreased differentiation [4]. Importantly, intracranial transplantation of transformed NPs into nonobese diabetic severe combined immunodeficient (NOD SCID) mice

Sheila K. Singh and Chitra Venugopal (eds.), *Brain Tumor Stem Cells: Methods and Protocols*, Methods in Molecular Biology, vol. 1869, https://doi.org/10.1007/978-1-4939-8805-1_12, © Springer Science+Business Media, LLC, part of Springer Nature 2019

revealed neuroectodermal tumors that displayed cellular and molecular features associated with the most aggressive types of medulloblastoma (MB) [4]. Subsequent studies from our laboratory further tested the utility of these models through genetic manipulation of neural derivatives from both normal and transformed hENs to study the effects of specific genes of interest on MB progression [5]. Recently, this model system has also been applied to other pediatric brain tumors such as diffuse intrinsic pontine glioma (DIPG) [6] as well as high-throughput drug screening platforms to identify drugs that selectively target putative cancer stem cell populations [7]. Collectively, these results underscore the value of hESCs and their derivatives as complementary model systems for investigating the molecular mechanisms contributing to tumor progression.

Here, we describe methods for generating NPs from hESCs in both adherent and tumorsphere culture in defined neural medium supplemented with basic fibroblast growth factor (bFGF) and epidermal growth factor (EGF). This includes a 2-Stage protocol for adapting hESCs to neural culture and subsequent passage as well as methods for investigating cell migration using the hanging drop/ 3-dimensional collagen assay. Procedures detailing genetic modification of hESC-derived NPs using lentiviral transduction for gain/ loss of function studies are outlined and include sorting strategies for selection of stably transduced cells in the absence of antibiotics. Protocols for conducting intracranial or intracerebellar transplantation of NPs and/or genetically modified NPs derived from hESCs into 5–7 week old NOD SCID mice using a stereotactic frame are described. Finally, we outline the procedures for perfusion and fixation of NOD SCID brains with associated tumor samples for histopathological analysis or the collection and sorting of cells for secondary injection into NOD SCID mice as an in vivo assessment of self-renewal.

2   Materials

2.1   2-Stage Generation of Neural Precursor Cells from hESCs

1. Collagenase IV.

2. Poly-D-lysine or L-ornithine/laminin coated plates (6 and 24 well).

3. Ultra-low attachment multiwell plates (6 and 24 well).

4. Neural precursor (NP) medium: DMEM/F12 media, 1% N2 (100×) supplement, 2% B27 (50×) supplement, 20 ng/mL EGF human recombinant protein, 20 ng/mL bFGF human recombinant protein. Store at 4 °C away from direct light, as it is light sensitive, for 10 days (*see* **Note 1**).

5. 1× PBS.

6. Accutase dissociation agent.

7. Dimethyl Sulfoxide (DMSO) ultra-pure grade.

2.2 Hanging-Drop/3-Dimensional Collagen Migration Assay

1. Collagen type 1 (bovine).

2. 0.1 M NaOH (filter sterilized).

3. 10× DMEM.

4. Neural precursor (NP) medium: DMEM/F12 medium, 1% N2 (100×) supplement, 2% B27 (50×) supplement, 20 ng/mL EGF human recombinant protein, 20 ng/mL bFGF human recombinant protein. Store at 4 °C away from direct light, as it is light sensitive, for 10 days.

5. 48 multi-well regular tissue culture plates.

6. Petri dishes.

7. Ocular micrometer.

2.3 Lentiviral Transduction of hENs and Neoplastic hENs

1. Matrigel coated 24 multi-well plates.

2. Neural precursor (NP) medium: DMEM/F12 medium, 1% N2 (100×) supplement, 2% B27 (50×) supplement, 20 ng/mL EGF human recombinant protein, 20 ng/mL bFGF human recombinant protein. Store at 4 °C away from direct light, as it is light sensitive, for 10 days.

3. Accutase.

4. LentiORF lentiviral particles with TurboGFP selection marker and/or shRNAmir constructs with TurboGFP selection marker.

5. RFP positive control lentiviral particles.

6. Access to a fluorescent activated cell sorter (FACS).

7. 7AAD viability dye.

8. Fetal bovine serum (standard sterile-filtered).

9. Cell strainers (35 μm nylon mesh strainer with round-bottom tube).

10. Freezing medium; neural precursor medium (*see* above), 10% DMSO.

2.4 Intracranial and Intracerebellar Injections of Neoplastic NPs or Genetically Modified NPs

1. Dremel drill and #81 drill bit.

2. 10 μL Hamilton syringe and 32 g custom needle.

3. Stereotaxic mouse frame and bedding on which mouse lies.

4. Sterile drapes and sleeve for drill wire.

5. Sterile tuck towels and surgery towels.

6. Sterile recovery cage.

7. Sterile cotton swabs.

8. Small and large sterile gauze squares.

9. 1 cc and 3 cc syringes.

10. 250 mL saline bag.

11. Ultra-pure H_2O.

12. 70% EtOH.

13. #15 blade and blade holder.

14. 4.0 or 5.0 Dexon sutures.

15. Weigh scale.

16. Analgesic: Meloxicam 2 mg/kg.

17. Anaesthetic: Isoflurane (4% induction, 1–3% maintenance).

18. 18, 25, and 30-gauge needles.

19. 5–7 week old NOD SCID mice.

20. UltraMicroPump (UMP3) (one) with SYS-Micro4 controller.

21. Water-based heat pad (to be placed underneath recovery cage).

22. Eye gel.

23. Surgical tray: small bowl for 70% EtOH, large bowl for normal saline, needle driver, scalpel handle and blades, skin retractors, sterile green towels, scissors, two sets of tweezers, ear punch, gauze squares, forceps (blunt and sharp).

24. Microcentrifuge tubes with cells in sterile PBS on ice.

2.5 Perfusion and Fixation of NOD SCID Brains and Associated Tumor Samples

1. Stage and tape for securing the mouse.

2. Absorbent under pads.

3. Lab forceps set (tweezers, scissors, skull cracker).

4. Sterile saline or PBS.

5. Heparin (1000 USP units/mL).

6. 10% buffered formalin.

7. Anesthetic: Isoflurane (4% Isoflurane +1 L/min O_2).

8. 18/26/30 G needles.

9. 1 mL and 30 mL syringes.

10. 50 mL falcon tubes.

11. 70% ethanol.

2.6 Extraction and FACS of Primary Tumor Cells for Cell Culture and Secondary Tumor Injections

1. All materials included in Subheading 2.4.

2. CO_2 chamber.

3. Artificial cerebrospinal fluid (A-CSF): $1\times$ PBS, 5% fetal bovine serum (FBS).

4. Dissociation buffer: 9 mL DMEM F12, 450 μL FBS, 90 μL pen/strep (100×), 1 mL Accutase, 1 mL collagenase/Hyaluronidase (10×).

5. Neural precursor medium (*see* Subheading 2.1).

6. Poly-D-lysine and/or ultra-low attachment plates (6 and 24 wells).

7. HLA-ABC, APC conjugated antibody.

8. 7AAD viability dye.

3   Methods

3.1   2-Stage Generation of Neural Precursor Cells from hESCs

Stage 1

1. Dissociate normal hESCs (H9) [8] and transformed hESCs grown on Matrigel in mTESR ™ as previously described [5] for 5 min in collagenase IV.

2. Scrape cultures using a 5 mL or 10 mL pipette to generate small aggregates (*see* **Note 2**), gently triturate 5× and then replate aggregates as follows.

Both adherent and suspension cultures are grown in neural precursor (NP) medium.

(a) For adherent culture: 6-well or 24-well poly-D-lysine/laminin coated plates.

(b) For suspension culture (tumorspheres/neurospheres): 6-well or 24-well ultra-low attachment plates.

3. Grow adherent and suspension cultures for up 7 days, until they reach confluency with a media change every other day (*see* **Note 3**).

Stage 2

4. Once the cells reach confluency, rinse in PBS and dissociate the adherent and suspension cultures into single cells using Accutase for 5–10 min (5 min for adherent cultures; 10 min for suspension cultures).

5. Collect cells in 15 mL conical tubes and gently spin down for 5 min at 201 × *g*. Aspirate remaining Accutase/PBS and replate as follows (*see* **Note 4**):

(a) For adherent cultures: replate in 24- or 6-well poly-D-lysine/laminin coated plates.

(b) For suspension cultures: Replate into 24- or 6-well ultra-low attachment plates in NP medium to facilitate tumorsphere/neurosphere formation over 5–7 days.

6. For maintenance and passage, change medium every 3 days. After 5–7 days, dissociate adherent cultures and neurospheres/tumorspheres into single cells using Accutase, gently spin and replate as stated above. NPs can be analyzed at this point for expression of neural stem/progenitor cell markers such as nestin by flow cytometry or western blot (*see* **Note 5**).

7. For long-term storage, dissociate adherent and/or neurosphere/tumorsphere cultures to a single cell suspension using Accutase. Once cell pellets have been washed with PBS, gently resuspend the pellet with NP medium supplemented with 10% DMSO. Aliquot 1 mL per labeled cryotube and quickly transfer the cells to cryo-containers for overnight storage at −80 °C. Transfer frozen cells to liquid nitrogen the following day. One full well of adherent or neurosphere cell culture from a 6 multi-well plate can be split into 2–3 frozen aliquots (1 mL per aliquot).

3.2 Hanging-Drop/3-Dimensional Collagen Migration Assay

3.2.1 Generation of Hanging Drops

1. Wash and dissociate cells with Accutase to obtain a single cell suspension.

2. Resuspend the cells in a small volume and do a live cell count.

3. Prepare a cell suspension of 1.25×10^6 cells/mL as this will yield 2.5×10^4 cells/20 μL drop.

4. Pipette 20 μL "drops" onto the underside of a lid of a petri dish.

5. Fill the bottom of the petri dish with 5 mL PBS to prevent the drops from drying out.

6. Slowly flip the lid back onto the dish and incubate the drops for up to 4 days (*see* **Note 6**).

7. Once the drops are composed of compact cell aggregates, prepare fresh collagen as described below.

3.2.2 Implanting Spheroids

The following steps should all be conducted on ice as the collagen will solidify at room temperature.

1. If using a new vial of collagen type 1, transfer entire content of vial to a 15 mL conical tube and label with date and content. This can be returned to the 4 °C storage for future use.

2. Add appropriate volume of collagen type 1 and 10× DMEM, mix gently, trying not to make bubbles as this solution is viscous. The solution should be bright yellow.

3. Adjust the pH of the solution using 0.1 M NaOH and mix thoroughly. The solution will become bright pink. If not, slowly add more NaOH (100 μL at a time).

4. Place the 48-well plate on ice and aliquot 110 μL collagen solution into each well. Gently pipette one drop (using a

p1000) containing your spheroid and deposit into the bottom of a collagen-coated well. Try not to introduce bubbles (*see* **Note 7**). Repeat with remaining spheroids.

5. Incubate the implanted spheroids for a *minimum* of 30 min at 37 °C (not more than 1 h). The collagen will solidify and become translucent.

6. Overlay the spheroids with another 100 µL of collagen by gently pipetting it alongside the well walls, so as not to disturb the first layer of collagen.

7. Incubate the spheroids for another 30 min at 37 °C.

8. Once you are confident that the entire collagen sandwich is solidified, add 200 µL of medium on top.

9. Measure the width and length of each implanted spheroid that same day (day 0).

10. Take measurements each day for 3 days.

11. Total migration is calculated by subtracting day 0 measurements (average diameter day 0) from day 1–3 final measurements (average diameter at days 1, 2, and 3).

3.3 Lentiviral Transduction of hENs and Neoplastic hENs

3.3.1 Lentiviral Infection

1. Plate 5×10^4 hEN cells per well of a 24- poly-D-lysine/laminin plate in NP medium [5]. Incubate overnight.

2. The following day, proceed to treat the cells with the lentiviral particles, at the desired MOI, as per the manufacturer's guidelines (*see* **Note 8**).

3. Feed the cells with NP medium 6–8 h following lentiviral infection (*see* **Note 9**).

4. Culture the cells for 48–96 h before proceeding to fluorescent activated cell sorting (FACS). During this time, GFP/RFP can be visualized (*see* **Note 10**).

3.3.2 Selection of Gain/Loss of Function Cells

5. Controls for the sorting process include unstained/untreated cells, 7AAD only cells, and GFP/RFP+ only cells (*see* **Note 11**).

6. Dissociate the cells using Accutase as previously described (as described in Subheading 3.1, stage 2), and collect in a 15 mL conical tube.

7. Gently wash the cell clumps with 1× PBS, spin for 1 min at $201 \times g$, and aspirate the supernatant.

8. Resuspend the cells in 1× PBS + 0.5%FBS (*see* **Note 12**) and filter through 35 µm cell strainers to ensure a single cell suspension (*see* **Note 13**).

9. Count live cells to estimate the concentration of cells to be sorted. The sorting process will select for GFP/RFP only expressing cells

10. Add 2 μL 7AAD (viability dye) for every 2 × 105 cells to be sorted as well as the appropriate controls. Add directly to the filtered cells 10 min before sorting (*see* **Note 14**).

11. As soon as the sorting process is complete, wash the cells with 1 × PBS.

12. Plate them back in 24-well poly-D-lysine/laminin plates.

13. The resulting cultures should be entirely GFP/RFP positive.

14. Once the cultures have recovered from the sorting process and are expanded, make frozen stocks.

15. To freeze gain/loss of function hENs, dissociate cells with Accutase, wash and spin the cells. Aspirate the supernatant and add freezing media. Aliquot cells into cryotubes (*see* **Note 15**). Freeze in −80 °C overnight and store in liquid nitrogen for long term.

16. Test the resulting gain/loss of function via SDS-PAGE and western blotting as well as qPCR to validate the stable lines before proceeding with further experiments. An example of lentiviral-transduced hENs with associated controls as well as a representative migration assay is shown in Fig. 1.

3.4 Intracranial and Intracerebellar Injections of Neoplastic NPs or Genetically Modified NPs

1. Spray down all equipment and work surfaces with 70% alcohol to sterilize.

2. Place stereotaxic frame and bedding in center of surgical stage. Set recovery cage (with water-based heat pads placed underneath the cage for additional warmth) and weigh scale in close proximity. Place drill with drill bit inserted in front left corner, plugged in, draped and covered in sterile towels on work surface. Mark drill bit with pen at 5 mm depth. Place green surgical towel in center of work surface along with all surgical instruments there.

3. Weigh first animal and transfer to induction chamber (induction with 4% Isoflurane +1 L/min O_2) followed by maintenance (1–3% isoflurane) via mask.

4. Quickly transfer mouse from the induction chamber over to the stereotaxic stage.

5. Place front top teeth onto the palate bar and slide gas anesthesia mask over nose. Secure the mask and ensure anesthetic is between 1 and 3% Isoflurane.

6. Add gel to the eyes to prevent drying.

7. Immobilize mouse in frame by inserting ear holders into external auditory canals and gently roll mouse away from the user onto posterior/far ear holder, which remains fixed. Gently advance anterior/near ear holder into mouse ear and roll mouse back toward the user. The head is now fixed so that

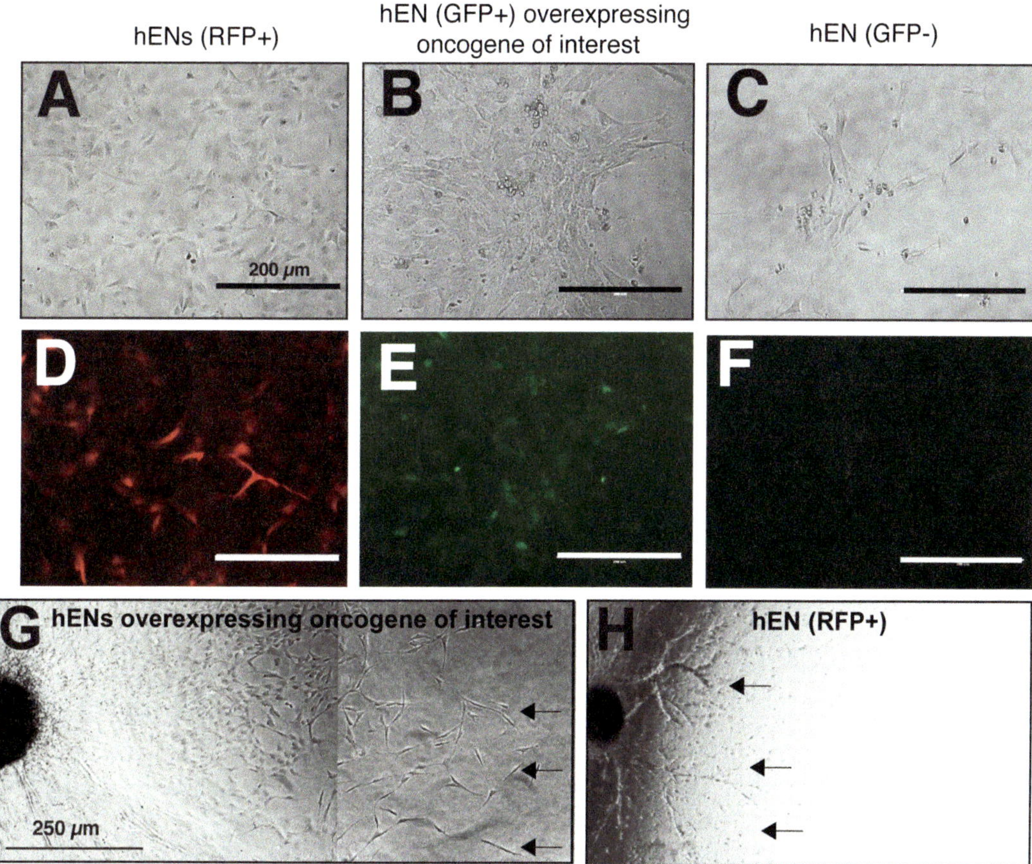

Fig. 1 Representative images of hENs stably overexpressing lentiviral constructs consisting of either RFP or GFP/oncogene of interest. (**a–f**). Long-term stable expression of RFP and GFP in sorted cell populations. Brightfield (**a–c**) and fluorescent (**d–f**) images of sorted RFP+ hEN (**a, d**), hENs overexpressing an oncogene of interest (GFP+) (**b, e**), and hEN (GFP-) (**c, f**) following 2 months extended culture. Scale bar: 200 µm. (**g–h**) Representative migration assays comparing sorted hENs overexpressing oncogene of interest (**g**) with hENs (RFP+ controls). Cells were cultured as hanging drop aggregates for 3 days and then placed into 3-dimensional type I collagen gels to follow migration over 72 h. Arrows denote migration front

the nose will not droop. Double check that the animal's top teeth are still securely set in the mask.

8. Check the pedal reflex to ensure animal has reached appropriate depth of anesthesia before proceeding (*see* **Note 16**).

9. Inject the dose of analgesic (meloxicam) 0.1 mL per animal (2 mg/kg, 0.5 mg/mL meloxicam solution further diluted with saline depending on the weight of the animal) subcutaneously prior to surgery. This will ensure the drug is active upon completion of the surgery when the animal is recovering.

10. Clip hair over the incision site.

11. Prepare the animal for surgery using an antiseptic surgical scrub. The skin is washed with 70% ethanol. Work in a circular motion, starting at the proposed incision site and working outward. Discard cotton swab once non-clipped skin is reached. Repeat two more times. If contamination occurs, the scrub should be repeated (*see* **Note 17**).

12. Make vertical midline incision through skin and scalp layers to periosteum (*see* **Note 18**). The landmark for the coronal suture is the anterior ear fold, so the incision should be 1.5 cm, equidistant on either side of the anterior ear fold. Clear periosteum with knife blade or small gauze square, and place skin retractors at the incision.

    (a) Landmark for the frontal lobe injection: 2–3 mm anterior to the coronal suture, 3 mm lateral to the midline, 5 mm deep.

    (b) Landmark for cerebellar injection: 1.5 mm posterior from the center of the lambdoid suture toward cerebellum and 2.5 mm deep.

13. Attach the drill to holder and the drill to the frame. Position drill at correct location (**a or b as stated above in step 12**) and ensure that it is perpendicular to mouse skull (*see* **Note 19**). Drill skull to appropriate depth.

14. Clean burr hole with EtOH soaked swab after drilling and ensure bleeding has stopped (*see* **Note 20**).

15. Prepare Hamilton syringe (previously rinsed with EtOH and sterile H_2O) and draw up 5–10 µL of cells (genetically modified hENs or neoplastic hENs) with 30 gauge needle. Attach the syringe to the UltraMicroPump, stereotactic frame and syringe holder. Insert the needle into the burr hole to appropriate depth (indicated above) based on site of injection.

16. Depress syringe over the adjusted time and speed on the UltraMicroPump (*see* **Note 21**).

17. Once the injection is done, leave needle in place for approximately 2 min, manually bring the needle out 1 mm, then wait another 2 min, then slowly remove the needle completely.

18. Clean injection site with EtOH soaked cotton swab and hold with gentle pressure for 30 s to prevent any bleeding.

19. Close incision in one layer interrupted PDS or Maxon stitches (3–4 will suffice).

20. Gently clean wound with 70% EtOH and remove mouse from frame.

21. Notch the mouse's ear to label and inject 2–3 mL normal saline to ensure proper hydration during recovery.

22. Place mouse on side in recovery cage on sterile gauze.

23. Repeat with subsequent mice.

3.5 Perfusion and Fixation of NOD SCID Brains and Associated Tumor Samples

1. Clean biosafety cabinet before setting up.

2. Build ramp and cover with absorbent pad.

3. Take out lab forceps, clean with EtOH and place in hood.

4. Anesthetize mice using isoflurane to surgical depth and test using pedal reflex. The animal can overdose since it will be sacrificed.

5. Use tape to secure the mouse to the ramp.

6. Spray body with EtOH.

7. Cut open abdomen horizontally into squares to remove fur and skin.

8. Open cavity and diaphragm to reveal heart and internal organs. Take care not to puncture the heart. The mouse will gasp and this is normal as you are affecting breathing.

9. Cut open the rib cage. With the help of the tweezers, get a hold of the heart and inject 0.50 mL of heparin at apex of heart to thin blood.

10. Wait 10 heart beats.

11. Grab needle and push into the apex of heart—attached to syringe filled with 15–20 mL saline, cut auricle, and inject until the blood has cleared and only clear fluid comes out.

12. Take out the saline syringe (heart should still be pumping).

13. Slowly inject formalin (15–20 mL) with needle inserted at the apex of the heart. There will be muscular twitching throughout the body as formalin flows through. Let formalin out over a 2–3 min period (*see* **Note 22**).

14. Animal will be stiff afterward.

15. Cut off head and remove all fur, eyeballs, and muscles to reveal the skull and open the spinal column by breaking the surrounding bones.

16. Remove skull using the pliers and extract the brain and spinal cord slowly from base of skull and spinal column (*see* **Note 23**).

17. Drop brain and tumor in 10% formalin for at least 72 h.

**3.6 Preparation
of Primary Tumor
Samples for FACS
and Subsequent
Culture or Secondary
Tumor Injection**

*3.6.1 For Samples to Be
Cultured*

1. Sacrifice the mouse using CO_2 chamber.

2. Cut off the head.

3. Crack open the skull and remove the brain as described in Subheading 3.4 (*see* **Note 24**).

4. Transfer the brain into prepared A-CSF on ice. Samples are placed in a small amount of A-CSF in a culture dish and cut/minced into several pieces.

5. Transfer into 15 mL conical tube with dissociation buffer. Incubate at 37 °C rocking for 60 min (*see* **Note 25**).

6. Once the sample is completely dissociated, wash it with PBS. Spin at $453 \times g$ for 3 min.

7. Aspirate the supernatant and resuspend the sample in A-CSF. Filter through a 35 μm cell filter to ensure single cell suspension.

8. Count the cells and proceed to staining with HLA conjugated antibody for FACS sorting of human cells (*see* **Note 26**).

9. Add appropriate amount of 7AAD ($2 \mu L/2 \times 10^5$ cells) viability dye to the sample and controls 10 min prior to sorting.

10. Once the cells have been sorted for HLA+/7AAD-, wash with PBS and spin for 3 min at $453 \times g$.

11. Aspirate the supernatant and resuspend sample in NP medium. Plate cells into either neurosphere or adherent assay conditions as indicated in Subheading 3.1.

*3.6.2 For Secondary
Tumor Injections*

1. Follow instructions from 1 to 10 as indicated above for samples used to re-culture.

2. Aspirate the supernatant and resuspend the cells in a small volume of PBS and count the total number of live cells.

3. Prepare the cells for injections (10 μL per injection) based on the chosen total number of live cells. The number of cells injected will vary depending on the goals of the experiment.

4. Transfer cells onto ice and bring to surgery room for intracranial or intracerebellar injections as described in Subheading 3.3.

4 Notes

1. This medium was utilized by Pollard et al. [9] to grow glioma stem cell lines in adherent culture and we have adopted these conditions for our hESC-derived neural precursors.

2. hESCs and transformed hESCs are very sensitive to dissociation. Immediately dissociating into a single cell suspension prior to replating in neural culture has been shown to be very

harsh on the cells resulting in significant cell death. Transitioning through aggregate culture on the first passage helps to sustain a higher viability.

3. Transition of hESCs into neural culture often results in a significant amount of cell death. Replenishing the media every other day or even daily helps to sustain cell viability during this transition.

4. For both adherent and tumorsphere/neurosphere cultures, replate 5×10^4 cells and 2×10^5 cells in 24-well and 6-well plates respectively, as we have found this to be the optimal cell number to sustain growth in both conditions. The suspension cultures at this point can now be called neurospheres/tumorspheres, as each sphere is presumably derived from a single cell and not an aggregate as long as the cells are plated at clonal density (≤ 10 cells/µL) [10].

5. For experiments involving tumorspheres/neurospheres in which sphere number is quantified, do not feed the cells after replating as agitation of any kind facilitates aggregation and thus the development of non-clonal spheres [11].

6. The number of days in hanging drop format varies and is dependent on how compact the spheroids become. This will vary between cell lines/primary cultures and can take from 2 to 4 days.

7. The introduction of small bubbles during the implanting stage can be alleviated using a fine needle (26 gauge) to remove/pop them.

8. We have found that an MOI of 0.3–1 is sufficient for stable integration while keeping a low copy number. Refer to the lentiviral datasheet and procedures provided by the manufacturer for further information.

9. LentiORF particles can be left on the cells overnight; however, if there is observed toxicity, remove the lentivirus mixture after 4–6 h treatment and replace with fresh growth medium.

10. GFP expression can be visualized under a fluorescent microscope as early as 24 h; however, waiting until 96 h to sort the cells will ensure a sufficient number post-sort for replating. As hENs do not respond well to antibiotic selection, it is necessary to sort by FACS for stable selection.

11. Each control should ideally consist of a minimum of 2×10^5 cells with the same number of cells in each tube. These controls include unstained cells (no lentiviral transduction), 7AAD (used as a viability marker for sorting of live cells only), and GFP/RFP+ cells only (no 7AAD). These controls must be utilized as part of proper sorting procedure.

12. FBS is included in the cell suspension to enhance viability during the sorting process. Controls are resuspended in 0.5 mL PBS/FBS and sort tubes are resuspended in 1 mL PBS/FBS.

13. It is very important to filter the cells as clumping will clog the FACS machine and halt the entire procedure.

14. In order to ensure purity of the sorted cells (GFP/RFP), we suggest selecting the highest GFP/RFP expressing cells (top 10–15% in terms of mean fluorescent intensity) to establish the stable gain/loss of function cell lines.

15. 1 well of 6-well plate can be frozen to 2 vials (1 mL/vial). Passage number remains the same when freezing and is increased to the next passage upon thawing [5].

16. It is crucial to constantly monitor the animals during surgery and adjust the Isoflurane/O_2 levels accordingly. Watch for consistent breathing and pink color throughout the body and tail. If changes in any of the vital signs are observed, reduce the isoflurane levels and wait for recovery before proceeding with the surgery.

17. The skin should be cleaned with sterile H_2O and 70% ethanol and as soon as the incision has been made, cleaning should be performed with only sterile H_2O.

18. It is important to make the incision at the right spot on the skull and not too far back, as this can cut the neck muscles and cause severe bleeding.

19. It is important to hold the skin with forceps while drilling the skull; otherwise, the skin will get caught in the drill and possibly injure the animal's head and eyes.

20. Once the hole has been drilled in the skull, it is necessary to stop the bleeding completely before starting the cell injection. This will prevent the cells from leaking out with the blood.

21. Settings used were based on the UltraMicroPump as follows: Infuse (volume set): 200 nL, rate of injection: 60 nL/min. Note that this will slowly inject the cells into the brain but it is imperative to keep an eye on the syringe and stop the injection as soon as all liquid has been dispensed.

22. Muscular twitching indicates that fixation is working. During the perfusion, it is important to insert the needle exactly into the apex of the heart to enable saline and formalin flushing throughout the entire body.

23. Tumors on the surface are very sticky and care must be taken during removal.

24. The skull is much harder under these conditions and the sample will look a lot bloodier. Once the tumor has been

located, remove as much as the surrounding tissue as possible to isolate the section of interest. This will be helpful during sample dissociation for further processing.

25. The minced brain tissue should be pipetted very gently every 20 min during the incubation process to expedite dissociation. This can take anywhere from 60 min for small tumors to a few hours for a larger, more compact tumor. It is important to remove as much mouse brain tissue as possible while keeping the tumor intact.

26. The cells collected from the brain tumor will be a mixture of both mouse and human. Sorting of human cells is achieved using a human leukocyte antigen (HLA) antibody. Controls include unstained cells, HLA only stained cells and 7AAD only stained cells.

Conflict of Interest

The authors declare no conflict of interest.

References

1. Ben-David U, Benvenisty N (2011) The tumorigenicity of human embryonic and induced pluripotent stem cells. Nat Rev Cancer 11(4):268–277. https://doi.org/10.1038/nrc3034

2. Baker DE, Harrison NJ, Maltby E, Smith K, Moore HD, Shaw PJ, Heath PR, Holden H, Andrews PW (2007) Adaptation to culture of human embryonic stem cells and oncogenesis in vivo. Nat Biotechnol 25(2):207–215

3. Werbowetski-Ogilvie TE, Bosse M, Stewart M, Schnerch A, Ramos-Mejia V, Rouleau A, Wynder T, Smith MJ, Dingwall S, Carter T, Williams C, Harris C, Dolling J, Wynder C, Boreham D, Bhatia M (2009) Characterization of human embryonic stem cells with features of neoplastic progression. Nat Biotechnol 27 (1):91–97. https://doi.org/10.1038/nbt.1516

4. Werbowetski-Ogilvie TE, Morrison LC, Fiebig-Comyn A, Bhatia M (2012) In vivo generation of neural tumors from neoplastic pluripotent stem cells models early human pediatric brain tumor formation. Stem Cells 30(3):392–404. https://doi.org/10.1002/stem.1017

5. Kaur R, Aiken C, Morrison LC, Rao R, Del Bigio MR, Rampalli S, Werbowetski-Ogilvie T (2015) OTX2 exhibits cell-context-dependent effects on cellular and molecular properties of human embryonic neural precursors and medulloblastoma cells. Dis Model Mech 8 (10):1295–1309. https://doi.org/10.1242/dmm.020594

6. Funato K, Major T, Lewis PW, Allis CD, Tabar V (2014) Use of human embryonic stem cells to model pediatric gliomas with H3.3K27M histone mutation. Science 346 (6216):1529–1533. https://doi.org/10.1126/science.1253799

7. Sachlos E, Risueno RM, Laronde S, Shapovalova Z, Lee JH, Russell J, Malig M, McNicol JD, Fiebig-Comyn A, Graham M, Levadoux-Martin M, Lee JB, Giacomelli AO, Hassell JA, Fischer-Russell D, Trus MR, Foley R, Leber B, Xenocostas A, Brown ED, Collins TJ, Bhatia M (2012) Identification of drugs including a dopamine receptor antagonist that selectively target Cancer stem cells. Cell 149(6):1284–1297. https://doi.org/10.1016/j.cell.2012.03.049

8. Thomson JA, Itskovitz-Eldor J, Shapiro SS, Waknitz MA, Swiergiel JJ, Marshall VS, Jones JM (1998) Embryonic stem cell lines derived from human blastocysts. Science 282 (5391):1145–1147

9. Pollard SM, Yoshikawa K, Clarke ID, Danovi D, Stricker S, Russell R, Bayani J, Head R, Lee M, Bernstein M, Squire JA, Smith A, Dirks P (2009) Glioma stem cell

lines expanded in adherent culture have tumor-specific phenotypes and are suitable for chemical and genetic screens. Cell Stem Cell 4 (6):568–580. https://doi.org/10.1016/j.stem.2009.03.014

10. Pastrana E, Silva-Vargas V, Doetsch F (2011) Eyes wide open: a critical review of sphere-formation as an assay for stem cells. Cell Stem Cell 8(5):486–498. https://doi.org/10.1016/j.stem.2011.04.007

11. Coles-Takabe BL, Brain I, Purpura KA, Karpowicz P, Zandstra PW, Morshead CM, van der Kooy D (2008) Don't look: growing clonal versus nonclonal neural stem cell colonies. Stem Cells 26(11):2938–2944. https://doi.org/10.1634/stemcells.2008-0558

Chapter 13

Chromatin Immunoprecipitation (ChIP) Protocols for the Cancer and Developmental Biology Laboratory

Hunter McColl, Jamie L. Zagozewski, and David D. Eisenstat

Abstract

Chromatin immunoprecipitation assays permit the isolation and subsequent identification of genomic DNA (gDNA) fragments bound directly or indirectly to proteins of interest, including transcription factors, co-factors, or chromatin remodeling proteins. These isolated DNA fragments may include gene regulatory regions from enhancers, super-enhancers, promoters, and/or insulators. Cells of interest can be obtained from embryonic tissues at various developmental time points or cancer cells from patients or derived from model systems, including patient-derived xenotransplants and primary cancer stem cells and cell lines. ChIP variants include ChIP-reChIP to identify targets bound to different transcription factors or members of protein complexes or, alternatively, to characterize the histone modifications accompanying occupation of specific regulatory regions by the protein of interest, such as a transcription factor. Subsequent analysis of ChIP experiments includes standard PCR, quantitative PCR (qPCR), and ChIPseq.

Key words Chromatin immunoprecipitation, DNA fragmentation, DNA crosslink, Transcription factor, Histone modifications

1 Introduction

The foundation of chromatin immunoprecipitation (ChIP) is the ability to reversibly crosslink protein to DNA using formaldehyde [1] or other cross-linking methods. The ChIP technique was rapidly adapted for studies of histone modifications on chromatin, using gel electrophoresis and PCR to analyze the chromatin [2–4]. As crosslinking techniques became more refined it became possible to use the DNA:Protein complexes in immunoprecipitation assays [5, 6].

The modern ChIP assay is used to determine the occupation of chromatin regions by specific proteins of interest [7]. Typically, these proteins will be transcription factors (such as those encoded by homeobox, basic HLH, or forkhead genes), co-factors or

---

Hunter McColl and Jamie L. Zagozewski contributed equally to this work.

Sheila K. Singh and Chitra Venugopal (eds.), *Brain Tumor Stem Cells: Methods and Protocols*, Methods in Molecular Biology, vol. 1869, https://doi.org/10.1007/978-1-4939-8805-1_13, © Springer Science+Business Media, LLC, part of Springer Nature 2019

chromatin remodeling proteins, as these interact with chromatin to influence gene expression [8]. The basic outline of the ChIP assay is to isolate the cells/tissue of interest and reversibly crosslink chromatin to any proteins and complexes that interact with the chromatin structure. Once crosslinked, the chromatin is fragmented and extracted while still bound to these proteins. Introduction of an antibody specific to the protein of interest allows for the specific selection of chromatin that is interacting with this protein. Once isolated, the crosslinks are reversed and the chromatin is separated from the protein, usually by protease digestion. The isolated chromatin can now be used in other experiments or analyzed to determine which regions are enriched due to interactions with the protein of interest.

2   Materials

2.1   Reagents

1. Bovine Serum Albumin—(BSA).

2. Protease Inhibitor Cocktail: (PIC) can be frozen at −20 °C indefinitely at 25×. Acquired from Roche (Catalogue #11697498001).

3. UltraLink Protein A/G beads—Acquired from Pierce (Catalogue #53132) (*see* **Note 1**).

4. tRNA—Fisher (Catalogue # AM719).

5. RNaseA—Thermo Scientific (Catalogue # EN0531).

6. Proteinase K—Thermo Scientific (Catalogue # EO0491).

7. Micrococcal Nuclease—Takaraka (Catalogue # 2910A).

2.2   Buffers

1. 1% Paraformaldehyde (PFA): Dilute from a 4% solution as a 1% solution must be made fresh but 4% can be made in advance and frozen at −20 °C until needed. For 500 mL: 20 g PFA + 25 mL of 20× PBS, raise to 500 mL with ddH$_2$O. Dissolve until clear at 65C, pH to 7.5 and filter sterilize.

2. 1 M NaCl: For a 1 L stock prepare 58.44 g powdered NaCl in 800 mL ddH$_2$O, raise to 1 L.

3. 5 M NaCl: Prepare a smaller volume—5.84 g powdered NaCl in 15 mL ddH$_2$O, raise to 20 mL.

4. 0.5 M EDTA: For a 1 L stock, prepare 186.1 g dry EDTA in 800 mL ddH$_2$O, pH to 8.0 and raise to 1 L with ddH$_2$O.

5. 1 M Tris–HCl (pH 6.5)—for a 1 L stock prepare 121.14 g Tris in 800 mL ddH$_2$O, pH to 6.5 with HCl, raise volume to 1 L.

6. Dilute Buffer: 0.01% SDS, 1.1% Triton X-100, 1.2 mM EDTA, 16.7 mM Tris–HCl (pH 8.1), 167 mM NaCl.

7. Lysis Buffer: 1% SDS, 10 mM Tris–HCl (pH 8.1), 10 mM EDTA—prepare fresh; do not store.

8. LiCl Buffer: 0.25 M LiCl, 1% deoxycholate, 1 mM EDTA, 10 mM Tris–HCl (pH 8.1), 1% NP-40.

9. High Salt Buffer: 0.1%SDS, 1% Triton X-100, 2 mM EDTA, 20 mM Tris–HCl (pH 8.1), 500 mM NaCl.

10. Low Salt Buffer: 0.1%SDS, 1% Triton X-100, 2 mM EDTA, 20 mM Tris–HCl (pH 8.1), 150 mM NaCl.

11. Elution Buffer: 1% SDS, 0.1 M $NaHCO_3$—prepare fresh; do not store.

12. Re-ChIP elution buffer: 1× TE, 2% SDS, 15 mM DTT—prepare fresh; do not store.

13. TE Buffer pH 8—Thermo Scientific (Catalogue #AM9849).

2.3 Equipment

1. Temperature controlled tabletop centrifuge.

2. #60 Sonic Dismembrator (Fisher Scientific)—set to 40% pulse strength, output control 4.

3. Desktop shaker.

4. Water Bath or Heat Block.

3 Methods

3.1 Standard ChIP Protocol

3.1.1 Day One

1. Harvest tissue of interest and prepare as desired; use the same amount of tissue (or cells) in each sample (*see* **Note 2**). For each experimental (+Ab) sample there should be an equivalent IgG control (i.e., 2 +Ab samples require 2 IgG control samples alongside). Alternatively, there should be an equivalent sample without specific antibody (−Ab) (*see* **Note 3**).

2. Wash with cold *1×PBS*.

3. Centrifuge at 400 × *g* at 4 °C for 5 min.

4. Gently dissociate cells by pipetting up and down using an appropriate pipet tip. Check to see that dissociation resulted in single cells—take care to not lyse any cells. Clumping of cells will result in unequal shearing and crosslinking across different samples; premature lysis will result in loss of chromatin.

5. Pellet cell suspension at 400 g at 4 °C for 5 min and remove the supernatant.

6. Fix cells with freshly made *1% PFA (pH 7.4) + 1× (final concentration) Protease Inhibitor Cocktail (PIC)*. Incubate at room temperature (RT) with rotation/rocking—incubation time depends on tissue type; examine current literature for the optimal time for the tissue of interest. For ChIP on cultured cells, fixation time in 1% PFA is typically 5–10 min,

followed by a 10 min Glycine quenching step (final concentration 0.125 M).

7. Spin fixed cells at 400 × g at 4 °C for 5 min and remove the supernatant. Keep on ice until beads are primed. *If necessary, cells can be frozen following the fixation and washing steps. If stopping at this stage ensure cells are kept at −80 °C until use and avoid freeze/thaw cycles.*

8. "Prime" beads as follows:

    (a) 60 µL of *Pierce UltraLink Protein A/G beads* (Cat#53132) for each sample including IgG or –Ab controls. Prepare in excess to ensure that there are sufficient beads for each tube (i.e., if preparing 4 samples, prime beads for 5 samples).

    (b) Add *Dilute buffer + 1× PIC up to 1 mL.*

    (c) Gently resuspend and centrifuge at RT 400 × g for 2 min.

    (d) Repeat wash and remove the supernatant.

    (e) Add X volume of *Dilute Buffer + 1× PIC* (X = original volume of added beads (60 µL × number of samples +1, i.e., 4 samples = 60 × 5 = 300 µL) = > this makes 50% beads.

    (f) Leave on ice until sonicated chromatin is ready.

    (g) If using *magnetic beads*, priming is not necessary; wash 3× in a 0.5% BSA/PBS solution prior to the addition of the antibody sample.

9. Wash cells 2× with cold *1× PBS* centrifuging after each wash. Resuspend cells in PBS and remove PFA traces.

10. Remove supernatant after last wash and add 400 µL of *Lysis Buffer + 1× PIC.*

11. Chromatin Fragmentation—two main methods can be utilized, mechanical or enzymatic.

    (a) *Mechanical*: Sonication of chromatin on ice 20 times at 15 s intervals with 30 s rest periods (40% pulse strength, output control = 4). For isolation of ~300 bp fragments we typically do 35 rounds per sample. Take care not to "foam" the lysis buffer as there is SDS present. Sample foaming will strongly hinder chromatin sheering. If foam is produced, wait until it subsides before continuing sonication. Foaming occurs when the sonicator tip is not sufficiently below the liquid surface.

    (b) *Enzymatic*: Micrococcal Nuclease (Mnase) is used to digest exposed chromatin. Exact buffers and mix will be included in the enzyme kit; however, addition of 1X PIC to the buffer mix is recommended. Use a 10,000 gel Units/reaction for each sample, incubate at 37 °C for

20 min. Mnase digestion on PFA crosslinked chromatin will enrich for mono-nucleosome populations of chromatin (~146 bp). This step also requires manual lysing of the cell post digestion. This can be performed at 35% pulse strength, in 5 s intervals for 30 s total sonication per sample.

12. Run 3–5 µL of sheared sample on a 1% agarose gel. Sheared chromatin will run as a streak with strong, bright bands showing level of fragmentation. Depending on the fragment size that is desired, it may be necessary to perform further fragmentation; remember, you can make the fragments smaller but not larger! Work carefully.

13. To each tube add 60 µL of "primed" beads (preclearing), rotate for 1 h at 4 °C.

14. Centrifuge $400 \times g$ at 4 °C for 5 min.

15. Transfer *supernatant* to a new tube, discard beads. *Save 25 µL of supernatant in a different tube labeled total input (TI).* TI contains a sample of the *total chromatin* from your sample and is necessary for analysis of your ChIP output.

16. To the collected supernatant, add *BSA and tRNA* to a final concentration of 500 µg/mL (the supernatant already contains PIC, so do not need to add).

17. Add 1–10 µg of Ab if the concentration is known, to tubes that designated as +Ab. Incubate at 4 °C overnight (O/N) with rotation (*see* **Note 4**). The IgG or –Ab tubes remain as is (do not add antibody).

18. Prime beads again as described previously (3.1, **step 8**); however, this time add *BSA and tRNA* to a final concentration of 500 µg/mL—this makes 50% beads. Incubate the primed beads at 4 °C O/N with rotation. This step is to prime beads to the same conditions that the sonicated chromatin samples are under.

3.1.2 Day Two

1. Add 60 µL of primed beads to each tube *including* the IgG or –Ab tubes. The antibody, and any complexes they are bound to, will be isolated by the beads.

2. Incubate at 4 °C O/N with rotation.

3. If using magnetic Dynabeads, only 2–4 h rotation with the antibody/beads is required.

3.1.3 Day Three

1. Before beginning, prepare *Elution buffer* and preheat to 65 °C.

2. Pellet beads at 600 g at 4 °C for 5 min.

3. Transfer the supernatant to a fresh tube; *keep beads and tube.* This supernatant contains all the unbound chromatin and will

be used as an Unbound Chromatin (UB) sample required for qPCR analysis and troubleshooting.

4. Wash beads with rotation at 4 °C as follows with 1 mL of each (resuspend, centrifuge to pellet, remove supernatant and proceed):

(a)	*Low salt wash buffer*	5 min
(b)	*High salt wash buffer*	30 min
(c)	*LiCl Buffer*	30 min
(d)	*TE buffer pH 8.0 (2×)*	5 min

5. After the last wash, add 250 µL of freshly prepared *Elution Buffer* (preheated to 65 °C).

6. Incubate for 15 min at RT with *agitation*; the best option is on a shaker set to medium/medium high. Alternatively, incubate and vortex every 5 min.

7. Centrifuge at 18,000 × *g* for 5 min at 4 °C.

8. Transfer the supernatant to a fresh tube, keeping beads.

9. Repeat elution **steps (5–7)** and combine two supernatants (final volume ~500 µL). Discard beads.

10. Add 10 µL of *RNaseA* and 25 µL of *5 M NaCl* to combined supernatants. This step digests any RNA or single stranded DNA in the sample, and reverses the formaldehyde crosslinking.

11. Incubate at 65 °C O/N.

3.1.4 Day Four

1. Add 10 µL of *0.5 M EDTA*, 20 µL *1 M Tris–HCl (pH 6.5)*, and 2 µL *Proteinase K*; incubate at 65 °C for 2 h. This digests proteins out of the eluant.

2. Clean up reaction using a PCR Purification Kit (Qiagen). Can also use standard phenol/chloroform/isoamylalcohol extraction method.

3. Analysis by PCR, qPCR, or high-throughput sequencing.

3.2 ChIP Variations

3.2.1 ChIP-reChIP

ChIP-reChIP is an advanced ChIP protocol that allows for the detection of chromatin specifically occupied by multiple proteins of interest. The first stage is a standard ChIP protocol (as above) concluding at the end of Day Three (DNA is eluted from beads, but crosslinking of proteins has not been reversed). A second ChIP procedure is then performed using the eluant from the first ChIP as the input. This assay selects for chromatin bound by two different proteins at the same time, the first round selecting for Protein A. The second round selects for any chromatin bound by protein B out of an input of chomatin already bound by protein A. One can also reverse the order of antibodies as an additional control (*see* below).

It is recommended that when starting this technique for the first time the experiment is run in tandem with one set using Antibody A in the first ChIP and the other set with Antibody B, alternating for the reChIP assay. This will help to determine which antibody order (A then B vs. B then A) provides the best enrichment.

<table>
<tr><td>

3.2.2 Day 1—Begin at
3.1.1, **Step 5**
of the Standard ChIP
Experiment as Above

</td><td>

1. Add 75 μL reChIP Elution Buffer +1× PIC to beads and agitate for 30 min @ 37 °C. reChIP buffer replaces the Elution buffer that will be used later; it elutes the chromatin from the beads. Alternatively, one can use the standard Elution buffer at this step.

2. Centrifuge at 400 × *g* @ 4 °C for 5 min; transfer the supernatant to a new tube.

3. Dilute to a volume of 1 mL with *Dilution Buffer and 1× PIC*; add *BSA and tRNA* to a final concentration of 500 μg/mL.

4. Save a 50 μL sample as INPUT A; now have UB (Unbound Chromatin), TI (Total Input), and Input A (input prior to antibody B).

5. Add 1–10 μg of *Antibody B* (experimentally predetermined), incubate O/N at 4 °C with rotation.

6. Prime beads as per 3.1.1, **step 8**, adding BSA and tRNA to a final concentration of 500 μg/mL. Incubate O/N at 4 °C with rotation.

</td></tr>
<tr><td>

3.2.3 Day Two
and Onward

</td><td>

1. Continue with the reChIP experiment as a standard ChIP, beginning at heading *3.1.2*.

2. Select the preferred analysis method and perform as you would for any ChIP. During analysis Input A can be used to show enrichment solely with antibody A, and is useful for controlling for enrichment on outputs reChIPped with antibody B.

</td></tr>
<tr><td>

3.3 ChIP Analysis

</td><td>

There are multiple options for analyzing the output of ChIP assays. Here, we discuss the advantages and limitation of several analysis options.

</td></tr>
<tr><td>

3.3.1 PCR Amplification

</td><td>

PCR amplification is the original analysis method. It involves a standard PCR reaction using the ChIP output as the DNA source. Evidence of immunoenrichment is determined by running the PCR reactions on an agarose gel and comparing the +Ab sample to either the IgG or −Ab sample. This analysis method is targeted to specific loci of interest using designed primers. It is important to run all standard controls (no template, positive control) for the PCR in addition to the controls for the Chip as you may end up seeing

</td></tr>
</table>

amplification of PCR products on all samples. This method typically uses only the outputs of the ChIP assay when presenting data, but the Unbound Chromatin samples saved in the protocol are used for controls and troubleshooting (e.g., no enrichment in the output means you should see amplification of your region of interest from the UB). This method of analysis is less frequently used unless there is a negative tissue control, with preference usually given to qPCR analysis. However, PCR amplification is commonly used as an initial check of your output (*see* **Note 5**).

3.3.2 ChIP qPCR

ChIP analysis by quantitative PCR is the most commonly used form of analyzing and presenting ChIP data. This method utilizes qPCR analysis to provide specific fold-enrichment of experimental outputs versus controls. The advantages it provides are a high level of sensitivity, allowing for accurate normalization and statistical analysis. There exist two different options for normalizing the data and determining fold-enrichment:

% Input Method

Signals from ChIP outputs are normalized to the Input controls, Unbound Chromatin (UB). Typically, UB are diluted to 1% of the original concentration prior to use in the qPCR assay; this dilution is important to perform for proper analysis. This gives an input fraction of 6.64 (given by the $[-\log2$ of $0.01]$). A standard qPCR reaction is performed (conditions depend on primers and amplification size), using 1–2 μL of the ChIP output DNA and the (diluted) UB as the templates for the reactions. The (averaged) input fraction Ct values are subtracted from the Input Fraction giving the Adjusted Input (in this case $[Ct_{\mathrm{av}} - 6.64 = \mathrm{Adjusted\ Input}]$). Finally, in order to determine fold enrichment perform the following calculation on each sample's average *Ct* value: $100 \times 2^{[adjusted\ input\ -\ average\ Ct\ ouput]}$. Using this calculation on the IgG control demonstrates the level of background chromatin the beads pick up, showing the true enrichment in the presence of the antibody.

Fold Enrichment Method

This method does not take Total Input into account, and instead directly compares signal to background (i.e., +Ab to IgG). The calculation is straightforward with two steps, a nonspecific adjustment and a fold-enrichment. Nonspecific adjustment is done by subtracting the *Ct* of the IgG sample from both itself and the +Ab samples. The values from this calculation are then normalized by undergoing a $2^{-\mathrm{ddCt}}$ calculation. Everything is relative to the IgG sample as $[Ct_{\mathrm{IgG}} - Ct_{\mathrm{IgG}} = 0,$ and $2^{-0} = 1]$, whereas the experimental sample undergoes $[Ct_{+\mathrm{Ab}} - Ct_{\mathrm{IgG}} = \mathrm{x},$ and $2^{-\mathrm{x}} = \mathrm{Fold\text{-}Enrichment}]$.

3.4 ChIP-Sequencing (ChIPseq)

ChIPseq is an analysis option that utilizes large-scale sequencing to avoid some of the limitations that standard ChIP experiments incur and has become an extremely popular technique since its inception [9]. Rather than analyzing the ChIP output via targeting genes through qPCR analysis, ChIPseq is used to identify any and all chromatin fragments isolated by ChIP. This allows for identification of novel targets of the proteins of interest and is unbiased. It does have several limitations that make it a less viable option than other analysis methods based on scale and cost, including accessing of bioinformatics expertise.

Quality of output is extremely important in samples prepared for ChIPseq, and purity is a very important factor when considering this. Prior to sending samples out for sequencing it is important to determine that your DNA quality is sufficiently high to warrant sequencing; otherwise your sequencing reads will be too low. Analysis on a Bioanalyzer (Agilent) or similar infrastructure is highly recommended. Other issues to consider are fragment size; as second generation sequencing relies on DNA alignments of fragments, there can be misrepresented enrichment of sequences sharing many similarities (*see* **Note 6**).

Refer to Fig. 1 for a schematic representation of the ChIP protocol using an antibody to a specific histone modification.

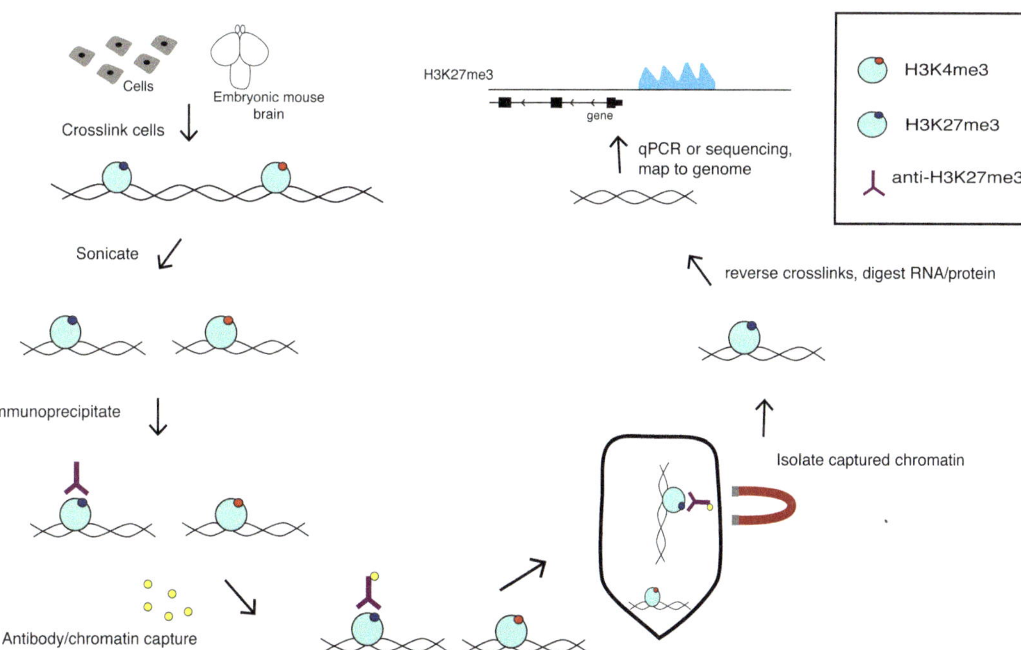

Fig. 1 Cells cultured in vitro or dissociated from tissues are first cross-linked to preserve protein/DNA interactions. The cross-linked chromatin is then mechanically or enzymatically sonicated to generate smaller chromatin fragments. To isolate the desired protein/DNA complexes, antibodies specifically raised against the protein of interest are added to the sonicated chromatin to immunoprecipitate chromatin complexes from the lysate. Antibody-bound chromatin complexes are then captured with magnetic or sepharose beads. Captured chromatin complexes are then isolated from the lysate with a magnet when using magnetic beads, or centrifugation when using sepharose beads. The enriched chromatin complexes are then eluted from the beads, cross-links reversed, and protein and RNA is digested from the solution. Following purification steps, regions of protein enrichment on the DNA are identified using PCR, qPCR, or high-throughput sequencing methods

4 Notes

1. Beads settle to the bottom of the bottle, make sure to resuspend before pipetting out. A popular alternative is magnetic Dynabeads—Thermo Scientific (Catalogue # 10002D) used with Magnetic Stand (Catalogue # 12321D).

2. The more tissue the better; yields from ChIP assays can be quite low, so use a *minimum of 30 mg of tissue or 1 × 10⁶ cells* for best quality. The amount of starting tissue will impact your assay, and it may be necessary to pool tissue samples together in order to obtain a suitable amount for the assay. A range from 1 to 20×10^6 cells often requires up to 30 mg of tissue. Although these quantities are generally considered suitable for a ChIP assay, the tissue quantity may be less if the protein of interest is localized to a specific anatomic region such as a cell layer or organ that can be dissected and triturated. Fixation times for each tissue type used will need to be determined keeping in mind that one wants to optimize protein-DNA crosslinking. Over- or under-fixed tissues will not yield successful ChIP outputs. When performing ChIP on cultured cells it is recommended to use at least 1×10^6 cells per sample.

3. *A mock sample* (Immunoglobulin G or IgG) is necessary when running a ChIP experiment in order to determine fold-enrichment over background. This mock sample is treated in the exact same manner as the experimental sample, but without addition of specific antibody. This negative control will demonstrate how much background enrichment is obtained through the experiment alone, as it is possible that the beads will bind and subsequently elute chromatin unbound by a specific antibody. A variant of using IgG is to run the ChIP protocol without the specific antibody to the protein of interest. A second control that is highly recommended is a *negative tissue control*, especially when studying transcription factors that are regionally expressed. When possible, it is recommended to run a ChIP assay in tandem with your experimental samples using tissues that do not express the protein of interest, such as those obtained from a knockout mouse model or alternatively from an embryonic organ that has little or no endogenous expression of this protein. This will help with troubleshooting and determining off-target or nonspecific binding of the specific antibody. However, negative tissue controls are usually not possible when studying histone modifications, which are ubiquitously expressed.

4. Most antibodies that are commercially available are not suitable for ChIP assays. ChIP grade antibodies require high concentrations and strong binding affinities. When possible, use

antibodies that have been published and shown to work in ChIP assays. If there are no published antibodies it may require some troubleshooting with multiple antibodies to find one that is suitable for the experiment. Monoclonal antibodies may bind with higher affinities or may be more specific for the protein of interest; however, the specific epitope targeted may not be accessible if the protein of interest interacts as part of a protein complex or has many posttranslational modifications. Polyclonal antibodies can bind with lower or higher affinity than monoclonal antibodies but may be less specific or result in higher background levels.

5. You may need to try different DNA Polymerases to find one that will amplify your sample well; we have found that high fidelity DNA polymerases such as Phusion (NEB) are the most reliable.

6. The most important limitation of ChIP is to understand that these experiments do not demonstrate direct binding of a protein to the chromatin; they show occupancy. Occupancy means that the protein of interest interacts with the chromatin, but not necessarily through specific binding. A protein that interacts with chromatin through a protein complex but without directly binding to the DNA itself will still show enrichment if used as the antibody target in a ChIP assay. This is due to the cross-linking of all proteins and complexes to the chromatin during the fixation step. Subsequent assays, such as electrophoretic mobility shift assays (gel shift) using nuclear extracts or recombinant proteins and labeled DNA fragments isolated from the ChIP experiment, may be required to further characterize whether binding is direct and/or specific.

References

1. Van Lente F, Jackson JF, Weintraub H (1975) Identification of specific crosslinked histones after treatment of chromatin with formaldehyde. Cell 5(1):45–50

2. Garner MM, Revzin A (1981) A gel electrophoresis method for quantifying the binding of proteins to specific DNA regions: application to components of the Escherichia coli lactose operon regulatory system. Nucleic Acids Res 9(13):3047–3060

3. Jackson V (1978) Studies on histone organization in the nucleosome using formaldehyde as a reversible cross-linking agent. Cell 15(3):945–954

4. Solomon MJ, Varshavsky A (1985) Formaldehyde-mediated DNA-protein cross-linking: a probe for in vivo chromatin structures. Proc Natl Acad Sci U S A 82(19):6470–6474

5. Orlando V, Strutt H, Paro R (1997) Analysis of chromatin structure by in vivo formaldehyde cross-linking. Methods 11(2):205–214

6. Solomon MJ, Larsen PL, Varshavsky A (1988) Mapping protein-DNA interactions in vivo with formaldehyde: evidence that histone H4 is retained on a highly transcribed gene. Cell 53(6):937–947

7. Orlando V (2000) Mapping chromosomal proteins in vivo by formaldehyde-crosslinked-chromatin immunoprecipitation. Trends Biochem Sci 25(3):99–104

8. Weinmann AS, Farnham PJ (2002) Identification of unknown target genes of human transcription factors using chromatin immunoprecipitation. Methods 26(1):37–47

9. Barski A, Cuddapah S, Cui K, Roh TY, Schones DE, Wang Z, Wei G, Chepelev I, Zhao K (2007) High-resolution profiling of histone methylations in the human genome. Cell 129(4):823–837

Chapter 14

EPH Profiling of BTIC Populations in Glioblastoma Multiforme Using CyTOF

Amy X. Hu, Jarrett J. Adams, Parvez Vora, Maleeha Qazi, Sheila K. Singh, Jason Moffat, and Sachdev S. Sidhu

Abstract

The ability to elucidate the phenotype of brain tumor initiating cell (BTIC) in the context of bulk tumor in glioblastoma multiforme (GBM) provides significant therapeutic benefits for therapeutic evaluation. For the identification of such an elusive and rare subpopulation of cells, a single cell analysis technology with deep profiling capabilities known as Mass Cytometry (CyTOF) can prove to be highly useful. CyTOF circumvents the spectral overlap limitations of traditional flow cytometry by replacing fluorophores with metal isotope tags, allowing the accurate detection of significantly more parameters at the same time. In this chapter, we demonstrate that synthetic antibodies can be conjugated with metal isotope tags for CyTOF analysis, resulting in the development of a highly tailored, custom multi-parameter panel. This toolset was used to stain patient-derived GBM cells, which was analyzed via CyTOF. Analysis software viSNE and SPADE were applied to study the co-expression patterns of the Eph Receptor (EphR) family and several putative BTIC markers in GBM, resulting in the identification of a distinct group of cells consistent with a BTIC subpopulation. This approach can be readily adapted to the detection of cancer stem-like cells in other cancer types.

Key words Mass cytometry (CyTOF), Synthetic antibodies, Cancer stem cells, Brain tumor initiating cells (BTICs), Glioblastoma and Eph receptors

---

1 Introduction

Cancer stem cells (CSCs) are subset of cells that are capable of seeding a tumor and differentiating to any other cell subtypes that comprise its tumor mass. These cells are of particular interest for therapeutic targeting as they represent the minimum cellular requirement for abhorrent growth [1, 2]. Though many studies have failed to observe subpopulations with these exact properties [3–5], the existence of CSCs is strongly supported by differential growth of engrafted tumor subpopulations. CSCs endowed with the tumor initiating properties that enable engraftment and nucleation of tumors in vivo are also known as tumor initiating cells

Sheila K. Singh and Chitra Venugopal (eds.), *Brain Tumor Stem Cells: Methods and Protocols*, Methods in Molecular Biology, vol. 1869, https://doi.org/10.1007/978-1-4939-8805-1_14, © Springer Science+Business Media, LLC, part of Springer Nature 2019

(TICs) whereas the cells that lack the ability to engraft are considered bulk, terminally differentiated progeny [2]. The highly proliferative and self-renewing properties of TICs have been implicated to drive tumor heterogeneity enabling a tumor mass to have a continuum of response to therapy [6]. TIC cells are rare subpopulations in passaged tissue cultures making of as little as <1% of the population [6]. Furthermore, they are dynamic and prone to rapid differentiation and redistribution once isolated and expanded in vitro or in vivo [7, 8].

Importantly, CSCs and TICs have been shown to display distinctive surface markers, which vary by tumor type enabling their detection [9]. Though a small component of total tumor mass, CSCs or TICs are the most optimal target-populations of therapy and therefore essential for clinical targeting. It is therefore necessary to create methodologies to study TIC phenotypes in the presence of bulk. Our ability to identify CSCs and TICs can pave the way for the development of efficient targeted therapies, with prolonged efficacy.

The dismal survival rate of Glioblastoma multiforme (GBM), the most common malignant primary brain cancer, is believed to be due to the existence of brain tumor initiating cells (BTICs) [10], which have the propensity to escape traditional treatments [11, 12]. Although many potential BTIC markers have been identified for GBM, such as CD133 [10], CD15 [13], and Bmi1 [14], their co-expression patterns have not been closely examined. Recently, two members of the erythropoietin hepatocellular carcinoma receptor (EphR) family, EphA2 and EphA3, have been identified as BTIC markers in GBM [15, 16]. Additionally, several other EphRs in the 14 member family, such as EphA7 [17] and EphB4 [18] have also been correlated with poor prognosis in GBM. Together these studies have identified this family of receptor tyrosine kinases as a consistent driver of tumorgenesis and critical markers of GBM BTICs. While obvious targets for cancer therapeutics, EphR biology is highly complex, resulting in the targeting of individuals EphRs to be ineffective [19]. Hypothesizing that the EphR family works cooperatively to drive oncogenesis in GBM, we set out to identify the complete expression patterns of the entire EphR family in GBM and determine how they further segregate BTIC subpopulations to study therapeutic targeting in heterologous patient-derived cell mixtures.

To address this hypothesis, a single cell analysis technology that is capable of measuring numerous cellular markers simultaneously is required. In an effort to evaluate the co-expression patterns of multiple BTIC markers and EphRs, CyTOF was employed for its deep profiling capabilities in complex biological systems [20]. This technology termed "mass cytometry" utilizes stable lanthanide metal isotope tagged antibodies instead of fluorophores, which allows the simultaneous detection of an unprecedented amount of parameters with negligible overlap at the single cell level [21, 22]. Although the metal isotope tagged antibodies required for the technology are commercially available for some cellular antigens, this collection is currently limited to mostly immunological targets. Fortunately, the availability of metal-labeling kits allows direct conjugation of metal isotope tags on the Fc region of any immunoglobulin (IgG) antibodies of interest. Although antibodies can be acquired through commercial means, these may be limited by availability, performance, or specificity. Alternatively, antibodies can be raised synthetically through phage display technologies [23]. Synthetic antibody design bypasses the limitations of natural immune systems in terms of diversity and functionality, allowing the development of exquisitely specific human antibodies toward virtually any target of interest that may have been otherwise unattainable [24]. Antibody fragments (Fabs), which are more easily produced via synthetic technologies than IgGs, are not easily labeled, but can be pre-clustered to an anti-Fab IgG labeled with a metal-isotope tag (Fig. 1), and subsequently used in CyTOF.

In order to test the efficiency of our synthetic antibodies in CyTOF, we developed a control experiment that can be processed in parallel with cell samples. In this test, antigens of interest are coated homogenously onto latex beads via passive adsorption and subsequently pooled (Fig. 2a). The resulting heterogeneous population of antigen bound beads is then stained with the same metal-isotope labeled antibody cocktail used for experimental cell samples for analysis via CytOF. If the antibodies used are effective toward their designated antigen and not cross-reactive, mutually exclusive subpopulations should be observed. In Fig. 2b, several members of

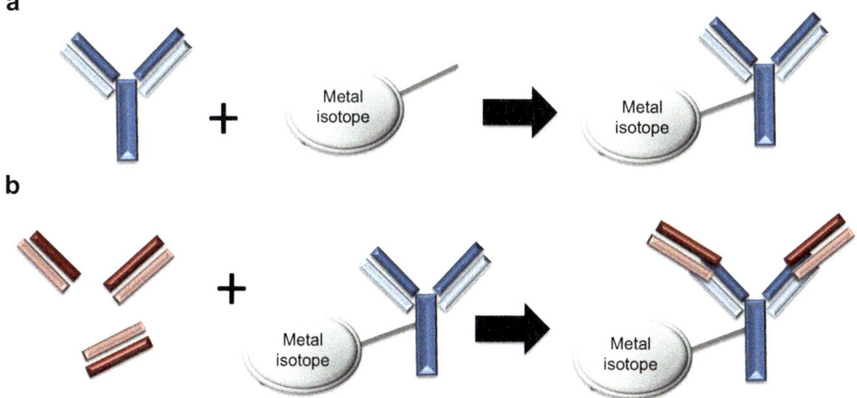

Fig. 1 Pre-clustering synthetic fabs to metal isotope labeled anti-Fab IgG. (**a**) an anti-Fab IgG is labeled with a metal isotope tag. Partial reduction of the IgG results in freed thiol groups that react with the maleimide group of the maxpar polymer (chelated with metal-isotopes), resulting in conjugation. (**b**) Fabs targeting an antigen of interest is incubated with metal-isotope labeled IgG at a 3:1 molar ratio. This produces a pre-clustered molecule that is specific for the antigen of interest, and is labeled with a metal-isotope for use in CyTOF

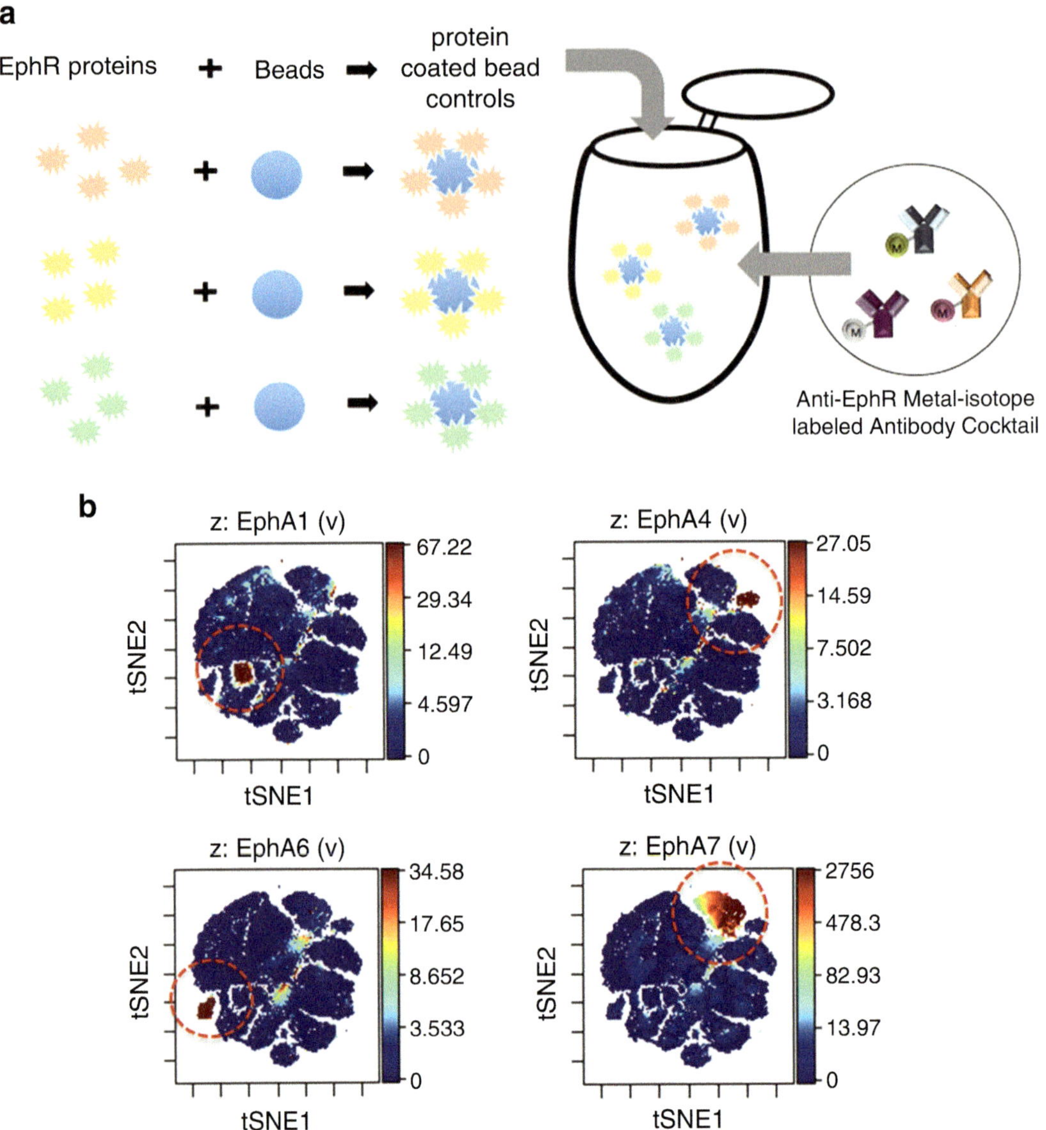

Fig. 2 Development of a positive control using antigen coated latex beads. (**a**) EphR proteins are homogenously coated onto aldehyde/amidine latex beads via passive adsorption. Different protein coated beads are then combined to simulate a cell population. The CyTOF antibody cocktail to be tested is then used to stain the coated beads mixture. (**b**) CyTOF and subsequent viSNE analysis graphs show mutually exclusive subpopulations with different EphR expression

the Eph Receptor (EphR) family are used to coat beads, which are then probed with recombinant EphR antibodies. The discrete, separate subpopulations observed using CyTOF computational analysis tool viSNE [25] demonstrate the high specificity of our synthetic antibodies toward even highly similar proteins (Fig. 2b). Conveniently, this strategy also simultaneously tests for metal-isotope labeling success. Thus, this control test presents an efficient way to assess the quality of diverse synthetic antibody cocktails that can be designed for use in CyTOF. By coupling the power of recombinant antibody technology with the potential of multi-parametric CyTOF, endless research possibilities may be realized.

We utilized a mass cytometry strategy with over 20 parameters in an effort to identify potential BTIC subpopulations in GBM. We included a comprehensive set of synthetic IgG antibodies toward the entire EphR family, which allow us the novel ability to systematically profile every EphR simultaneously on any population of human cells. In addition, we included commercial monoclonal antibodies toward putative GBM BTIC markers CD133 [10], CD15 [13], Bmil [14], ITGA6 [26], Sox2 [27], FoxG1 [28], as well as a non-BTIC differentiation marker MAP 2 [29]. Each of these antibodies were conjugated to distinct lanthanide tags, and used to probe patient-derived GBM cells. Resulting cells would then be labeled with metal isotope tags according to the proteins that they express. After the staining procedure, unbound antibodies were washed off, and an iridium-based DNA intercalator was added to allow the identification of individual cells. The labeled GBM cells were then analyzed via CyTOF, which utilizes time of flight mass spectrometry to distinguish the metal isotope tags by mass associated with each individual cell [21].

Upon analysis with CyTOF computational programs SPADE [30], we identified a discrete group of cells with heightened BTIC co-expression (Fig. 3), thus showing promise as a potential BTIC subpopulation. Due to the large number of surface markers we tested, we unexpectedly discovered other EphRs that are also co-expressed with the BTIC markers, such as EphA1 and EphB1 (Fig. 3). Thus, our method resulted in the novel discovery of additional surface markers that may also act as BTIC markers. The procedures outlined here can be easily adapted to assess a wide range of cell types for virtually any markers of interest.

2 Materials

2.1 Antibody Labeling

1. Maxpar® Antibody labeling kit: includes Maxpar® Polymer, lanthanide solutions, R-Buffer, C-Buffer, L-Buffer, W-Buffer.

2. Human monoclonal IgG1 antibodies, carrier-free (no BSA, gelatin, etc).

3. 0.5 M TCEP solution.

4. 3 kDa 500 μL V bottom centrifugal filter unit.

5. 50 kDa 500 μL V bottom centrifugal filter unit.

6. Aerosol barrier pipette tips.

7. PBS-based antibody stabilization solution.

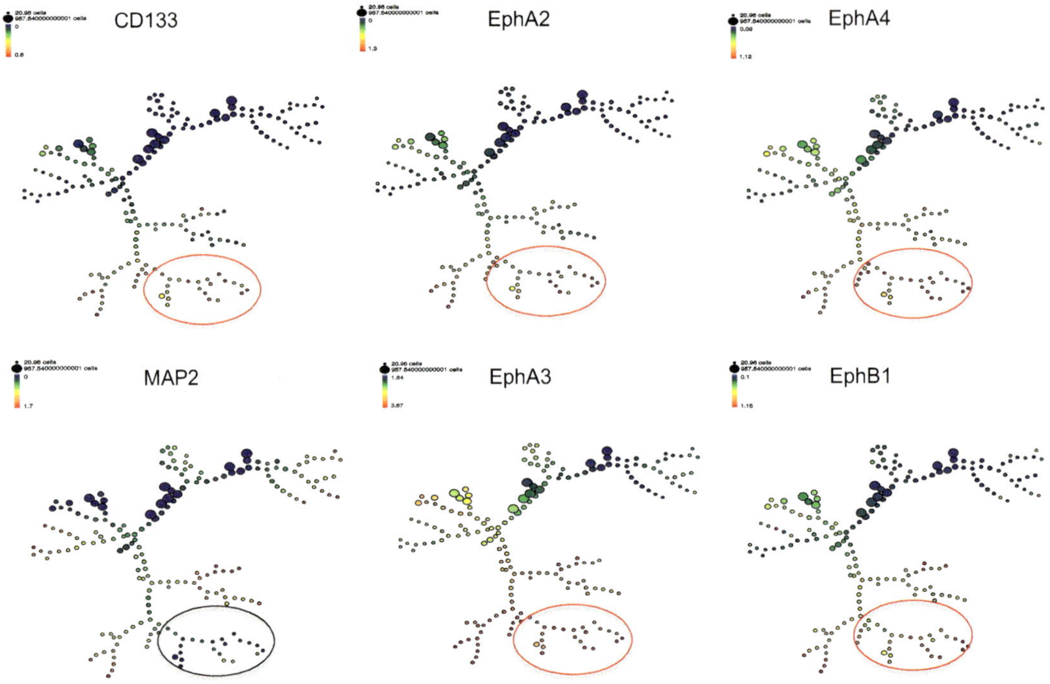

Fig. 3 SPADE analysis of glioblastoma cells shows distinct subpopulation of cells positive for BTIC markers EphA2, EphA3, negative for differentiation marker MAP 2 and positive for additional EphRs (EphA4, EphB1 shown). Metal-isotope labeled antibodies toward putative six BTIC markers, a differentiation marker and all EphRs were used to stain patient-derived glioblastoma cells. Metal signals were then analyzed via CyTOF. Computational technique SPADE clusters cells with similar expression profiles into groups visualized by circular nodes. The size of the node delineates number of cells contained within the group and the color of the node represents the median expression intensity (intensity scales vary between graphs). All cells displayed have been pre-gated for live singlets. Graphs shown are BTIC markers: CD133, EphA2, EphA3; differentiation marker MAP 2; other members of EphR family: EphA4, EphB1. Area circled in red represents a prospective BTIC subpopulation, which is approximately 5% of the total cells displayed. Unexpected co-expression of other EphRs in this BTIC subpopulation demonstrates the utility of CyTOF in identifying additional previously unknown BTIC markers

2.2 Working Solutions	1. PBS without magnesium or calcium (PBS−/−). 2. Stain Buffer: 1% bovine serum albumin, 0.1% sodium azide in PBS. 3. Deionized water (DI H$_2$O).

2.3 Cell Harvesting	1. Cell dissociation reagent appropriate for cell line of use:
	
	  (a) EDTA solution: 10 mM EDTA in PBS−/− OR Accutase®.
	
	  (b) Liberase™ TM solution: 0.2 U/mL Liberase™ TM in PBS−/− OR Accumax®.

2.4 Latex Beads Control	1. Aldehyde/Amidine latex beads 4% w/v, 0.06 μm.
	2. Antigen protein(s) of interest.
	3. MES buffer, 0.025 M, pH 6 (*see* **Note 1**).
	4. Deep well plate.

2.5 Viability and Surface Staining

1. Cell-ID Cisplatin-198Pt solution: dilute 5 mM stock 1000× in PBS−/− for use.

2. Surface stain cocktail (custom mix of lanthanide labeled antibodies toward cell surface targets diluted in stain buffer).

2.6 Intracellular and/or Intranuclear Staining (If Required)

1. Fixation/Permeabilization (Fix/Perm) buffer: commercial kit recommended (ex: BD Pharmingen Transcription Factor Buffer set).

2. Permwash buffer: included in commercial kits (recommended).

3. Intracellular and/or intranuclear cocktail (custom mix of lanthanide labeled antibodies toward intracellular or intranuclear cell targets diluted in permwash).

2.7 DNA Intercalation

1. Intercalation Buffer: 0.3% saponin, 1.6%–4% fresh paraformaldehyde in PBS.

2. Cell-ID Intercalator-Ir-125 μM. Dilute stock 1000× in intercalation buffer.

3. Test tubes with cell strainer snap cap.

2.8 CyTOF Acquisition and Analysis

1. CyTOF machine (ex: HELIOS).

2. EQ™ Four Element Calibration Beads.

3. Analysis platform compatible with fcs files (ex: Cytobank), preferably with computational analysis programs designed for CyTOF (ex: viSNE, SPADE).

---

3 Methods

Keep harvested cells and reagents on ice at all times unless otherwise specified.

3.1 Panel Design and Antibody Cocktail Preparation

1. To further avoid spectral mass—overlap due to mixed isotopes, select lanthanides tags for use with Fluidigm's Maxpar Panel Designer [31]. Pair up each antibody to be used with a different metal isotope tag.

2. Collect antibodies toward cellular markers of interest for labeling. The following options exist:

    (a) Commercially purchased ready-made metal isotope labeled antibodies.

   (b) Commercially purchased monoclonal IgGs (ideally validated by flow cytometry).

   (c) Synthetically raised monoclonal IgGs (validated by flow cytometry).

   (d) Synthetically raised monoclonal Fabs pre-clustered to an metal isotope labeled anti-Fab IgG:

- Select synthetic Fabs toward antigens of interest.

- Conjugate monoclonal anti-Fab IgGs with metal isotope tags (depending on how many fabs are used) according to **step 3**.

- Preincubate each distinct Fab with a different metal isotope tagged anti-Fab IgG for a minimum of 30 min at room temperature (RT). Considering a Fab:IgG molar ratio of 3:1, add equivalent amounts of Fab to IgG-Metal to ensure full saturation of both IgG arms.

- Pre-clustered Fab:anti-Fab IgG-Metal tag is ready for use.

   (e) Mix of above.

3. If applicable, conjugate antibodies to lanthanide metal tags through the use of Fluidigm's MAXPAR Antibody Labeling Kit [32] (*see* **Note 2**).

4. Antibodies can be titered on the CyTOF machine on cell lines of interest in order to determine optimal concentrations for use (*see* **Note 3**).

5. For each panel of antibodies to be used, make an antibody stain cocktail: add all metal isotope labeled antibodies at appropriate concentrations (as determined by titration) into a single tube, diluting in stain buffer (surface stain cocktail) or permwash (intracellular stain cocktail).

3.2 Harvesting Cultured Cells (ex: Patient-Derived Glioblastoma Cell Lines)

1. For cell line(s) of interest, grow up 1 to 3 million cells per test according to recommended culture conditions (*see* **Note 4**).

2. Remove culture media, and wash cells in flask with warm PBS. Remove supernatant.

3. Dissociate cells in pre-warmed media for 5 min at 37 °C. For adherently grown cells, add enough EDTA solution or Accutase to cover cells. For sphere forming cultures (ex: neurospheres), use Liberase™ solution or Accumax® (*see* **Note 5**).

4. Wash cells by topping up centrifuge tube with at least 10 mL PBS$-$/$-$: Spin 300 $\times$ *g* for 5 min and gently aspirate supernatant (*see* **Note 6**).

3.3 Antigen Coated Beads Control

1. Dilute proteins of interest in MES buffer to a concentration of 1 mg/mL (*see* **Note 1**).

2. Prepare latex beads by diluting 2.5 mL of 40 mg/mL stock solution with 10 mL MES buffer (*see* **Note 7**).

3. Wash beads by spinning down 3000 × g for 5 min. Remove the supernatant.

4. Resuspend pellet in 10 mL MES buffer and repeat wash step.

5. Repeat the wash step and resuspend final pellet in 5 mL MES buffer. The latex suspension is now approximately 20 mg/mL.

6. For each test, 5 μL of 20 mg/mL latex beads can be added to 5 μL of 1 mg/mL protein (*see* **Note 8**). Keep different proteins separate in individual wells of a deep well plate.

7. Incubate protein-beads mixture with gentle mixing at room temperature overnight.

8. Perform a wash step by adding 1 mL stain buffer and spinning at 3000 × g for 5 min to separate unbound protein from protein coated bead complexes. Remove the supernatant.

9. Resuspend the pellet in 1 mL stain buffer. Repeat **steps 6** and 7 twice for a total of 3 washes.

10. Resuspend to a final approximate concentration of approximately 5 μg/mL. Store at 4 °C until used. Do not freeze.

11. Combine equivalent amounts of different protein coated bead complexes to make desired cocktail at 5 μg/mL. This can now be stained with metal-isotope labeled antibodies as detailed in Subheading 3.5, followed directly by CyTOF acquisition as described in Subheading 3.8 (*see* **Note 9**).

3.4 Viability Stain

1. Filter cells through a cell strainer tube cap.

2. Count cells and resuspend up to 10 million/mL in PBS−/−.

3. Add Cell-ID Cisplatin to final concentration of 5 μM, ie 1 μL Cell-ID Cisplatin to 1 mL cell solution.

4. Mix well by vortexing gently (*see* **Note 10**). Incubate at RT for 5 min.

5. Quench staining by adding 5× the volume of stain buffer (i.e., 5 mL of stain buffer to 1 mL of cell solution).

6. Spin 300 × g for 5 min and aspirate the supernatant. Vortex gently to dissociate pellet.

3.5 Surface Stain

1. Aliquot 1–3 million cells per tube for staining.

2. Add 1–2 mL of stain buffer per tube, spin at 300 × g for 5 min, and remove the supernatant.

3. Vortex gently to resuspend pellet.

4. Add 50–100 μL of surface stain cocktail to cells at appropriate final concentration (*see* **Note 11**).

5. Mix well by vortexing gently.

6. Incubate at 4 °C for 20–30 min (*see* **Note 12**).

7. Wash by adding 1–2 mL stain buffer to tube, spinning and decanting supernatant. Resuspend pellet by gently vortexing.

8. Repeat for a total of 2 washes.

3.6 Intracellular/ Intranuclear Stain (If Applicable)

Fix and permeabilize cells for intracellular and/or intranuclear staining if needed. A commercial buffer set designed for flow cytometry staining (ex: BD Pharmingen Transcription Factor Buffer Set) is recommended. Follow instructions as dictated by commercially purchased set. In general:

1. Thoroughly disrupt pellet by vortexing gently while adding 1 mL of 1× Fix/Perm buffer dropwise.

2. Incubate samples in Fix/Perm solution at 4 °C for 40–50 min.

3. Wash cells 2× by adding 1–2 mL of 1× Perm/Wash buffer, spin $600 \times g$ 5 min, decant SN, vortex gently to loosen pellet (*see* **Note 13**).

4. Add 50–100 μL of intracellular stain cocktail in Perm/Wash buffer. Incubate at 4 °C for 40–50 min.

5. Wash cells: add 2 mL of 1× perm/wash, spin $600 \times g$ 5 min, decant SN, vortex gently to loosen pellet.

3.7 DNA Intercalation for Single Cell Detection (See Note 14)

1. Wash cells: add 2 mL of stain buffer, spin $600 \times g$ 5 min, decant SN, vortex gently to loosen pellet.

2. Dilute Cell-ID Intercalator-Ir 1000× in intercalator solution. Add 1 mL per tube and gently vortex.

3. Incubate for 1 h at RT or leave overnight at 4 °C.

4. Wash cells: add 1 mL of stain buffer, spin $600 \times g$ 5 min, decant SN, vortex gently to loosen pellet.

5. Repeat for a total of 2 washes with stain buffer.

3.8 CyTOF Acquisition

1. Add 2 mL of DI water and count cells. Wash by spinning $600 \times g$ for 5 min and discard SN.

2. Leave cells pelleted on ice or at 4 °C until ready to run on CyTOF machine (same day).

3. Shake EQ™ beads vigorously and dilute 1 part beads in 9 parts DI H$_2$O to make a 0.1× bead solution.

4. Immediately prior to acquisition, resuspend cell pellet to appropriate injection concentration with 0.1× EQ™ bead solution.

5. Filter cell solution through a cell strainer cap.

6. Acquire on CyTOF machine. An event rate of up to 1000 events/second can be used, or adjusted appropriately according to the nature of the cells under study.

3.9 Analysis

1. Apply EQ™ algorithm to normalize data if needed (included in CyTOF two software) [33].

2. Download normalized fcs files and upload to an analysis application or platform (ex: Cytobank).

3. Gate cells for event length, DNA content, singlets, EQ™ and viability.

4. Run CyTOF compatible analysis software such as viSNE or SPADE if desired.

4 Notes

1. The pH of the MES buffer should be close to the isoelectric point of the protein in order to maximize protein density on the particle surface. Do not use buffer systems with multivalent anions (such as PBS) prior to coating the latex beads as this may compromise colloidal stability [34].

2. Although the Maxpar antibody labeling kit indicates to dilute the labeled antibody to a concentration of 0.5 mg/mL in antibody stabilization buffer, we have found that with a starting maximum antibody concentration of 100 μg, the final resulting concentration is usually not much higher than 0.5 mg/mL. In order to maximize long-term stability, we recommend diluting the labeled antibody in an equal volume of antibody stabilization buffer regardless of its measured concentration, taking note of the diluted concentration. Working concentrations can then be calculated for each labeled antibody for use in CyTOF.

3. Titrate antibodies on CyTOF machine according to the same staining protocol used for the experiment. Even if the antibody has already been titered by flow cytometry, a newly conjugated antibody-metal isotope complex should ideally be tested on the CyTOF machine. Test a series of antibody dilutions on cell line of interest and determine the titration value at saturation.

4. If possible, attempt to avoid cell culture media with barium, as contamination with this metal will show up during CyTOF analysis. Similarly, avoid using glassware during all steps of the cell culture and staining procedures, as lead contamination from glassware will also be detected by CyTOF.

5. For adherent cell cultures, use the EDTA solution whenever possible in order to best preserve cell surface antigens for subsequent staining. Do NOT use Trypsin as this may cleave

proteins of interest. A slightly higher volume of the EDTA solution than the amount typically used with Trypsin may be required for effective dissociation. For strongly adherent cells that do not dissociate well with the EDTA solution, Accutase® may be used. For sphere forming cultures, Accumax® or diluted Liberase™ can be used for dissociation into single cells.

6. When washing cells, always ensure there is at least 1 mL of wash buffer before spinning down. To remove the supernatant after a wash, after ensuring a cell pellet is present, quickly invert the tube to dumb the fluid, and blot on a paper tower. However, prior to adding staining solution, it may be best to aspirate the supernatant so that there is less residual volume remaining.

7. A ratio of 5 mg of protein to 100 mg beads is optimal for coating. Thus, concentrations of beads and latex can be adjusted somewhat as long as this ratio is met. However, beads or protein concentrations that are too dilute or too concentrated may not be ideal for coating.

8. The latex beads should always be added second (into the protein solution). This order of addition helps to ensure the best coating of particles with the least amount of aggregation [34]. The protein is added at approximately 200% of what is required for a monolayer [34].

9. At this concentration, we have previously observed approximately a thousand protein-latex bead complexes (events) per µL. However, individual results may vary. Samples may be analyzed first via flow cytometry in order to estimate number of events. Approximately a million events can be used for a CyTOF staining experiment. With this amount, it is normal to not observe a pellet after centrifugation during the wash steps. By diluting in stain buffer, the BSA present can block any remaining reactive sites on the latex. Since these samples are beads rather than cells, Subheading 3.4–3.7 are not applicable. A slow event rate is recommended for initial acquisition. The latex beads selected for use are small enough to be completely vaporized by the CyTOF machinery [33].

10. To mix cells, pulse vortex gently. Do not pipette to mix as this will result in cell loss. Ensure the cell pellet is well resuspended (single cell suspension) before adding any staining solutions.

11. For cell staining, the final concentration of the antibody cocktail in the cell solution is of critical importance (ex: one might add 50 µL of the antibody cocktail at $2\times$ the final concentration as determined by titration to 50 µL of cell suspension). Thus, it is important to keep note of any residual volume after removal of the supernatant from the cell pellet after a wash.

12. For surface staining, keep cells on ice or at 4 °C to prevent internalization of cell surface receptors. The metabolic

inhibitor sodium azide, which is added to the stain buffer, also helps to prevent receptor internalization.

13. After cell fixation, all subsequent spin steps should be increased from $300 \times g$ to $600 \times g$, which will result in better cell recovery.

14. Even if cells were already fixed and permeabilized in previous steps, proceed with the intercalation buffer to ensure that cells are well prepared for DNA intercalation and subsequent washes in DI H_2O. Cells can be left at 4 °C in the intercalation solution for up to 48 h.

References

1. Reya T, Morrison SJ, Clarke MF, Weissman IL (2001) Stem cells, cancer, and cancer stem cells. Nature 414(6859):105–111

2. Zhou BB, Zhang H, Damelin M, Geles KG, Grindley JC, Dirks PB (2009) Tumor-initiating cells: challenges and opportunities for anticancer drug discovery. Nat Rev Drug Discov 8(10):806–823

3. Kelly PN, Dakic A, Adams JM, Nutt SL, Strasser A (2007) Tumor growth need not be driven by rare cancer stem cells. Science 317 (5836):337

4. Yoo MH, Hatfield DL (2008) The cancer stem cell theory: is it correct? Mol Cells 26(5):514

5. Rahman M, Deleyrolle L, Vedam-Mai V, Azari H, Abd-El-Barr M, Reynolds BA (2011) The cancer stem cell hypothesis: failures and pitfalls. Neurosurgery 68(2):531–545

6. Bao B, Ahmad A, Azmi AS, Ali S, Sarkar FH (2013) Overview of cancer stem cells (CSCs) and mechanisms of their regulation: implications for cancer therapy. Curr Protoc Pharmacol:14–25

7. Neuzil J, Stantic M, Zobalova R, Chladova J, Wang X, Prochazka L, Dong L, Andera L, Ralph SJ (2007) Tumour-initiating cells vs. cancer 'stem'cells and CD133: what's in the name? Biochem Biophys Res Commun 355(4):855–859

8. Wicha MS, Liu S, Dontu G (2006) Cancer stem cells: an old idea—a paradigm shift. Cancer Res 66(4):1883–1890

9. Gupta PB, Chaffer CL, Weinberg RA (2009) Cancer stem cells: mirage or reality? Nat Med 15(9):1010–1012

10. Singh SK, Hawkins C, Clarke ID et al (2004) Identification of human brain tumour initiating cells. Nature 432:396–401. https://doi.org/10.1038/nature03128

11. Chen J, Li Y, Yu TS et al (2012) A restricted cell population propagates glioblastoma growth after chemotherapy. Nature 488:522–526. https://doi.org/10.1038/nature11287

12. Liu G, Yuan X, Zeng Z et al (2006) Analysis of gene expression and chemoresistance of CD133+ cancer stem cells in glioblastoma. Mol Cancer 5:67. https://doi.org/10.1186/1476-4598-5-67

13. Mao X, Zhang X, Xue X et al (2009) Brain tumor stem-like cells identified by neural stem cell marker CD15. Transl Oncol 2(4):247–257

14. Abdouh M, Facchino S, Chatoo W et al (2009) BMI1 sustains human glioblastoma multiforme stem cell renewal. J Neurosci 29 (28):8884–8896

15. Binda E, Visioli A, Giani F et al (2012) The EphA2 receptor drives self-renewal and tumorigenicity in stem-like tumor-propagating cells from human glioblastomas. Cancer Cell 22:765–780. https://doi.org/10.1016/j.ccr.2012.11.005

16. Day BW, Stringer BW, Al-Ejeh F et al (2013) EphA3 maintains tumorigenicity and is a therapeutic target in glioblastoma multiforme. Cancer Cell 23:238–248. https://doi.org/10.1016/j.ccr.2013.01.007

17. Day BW, Stringer BW, Boyd AW (2014) Eph receptors as therapeutic targets in glioblastoma. Br J Cancer 111:1255–1261. https://doi.org/10.1038/bjc.2014.73

18. Pasquale EB (2010) Eph receptors and ephrins in cancer: bidirectional signalling and beyond. Nat Rev Cancer 10(3):165–180

19. Lu-Emerson C, Norden AD, Drappatz J, Quant EC, Beroukhim R, Ciampa AS, Doherty LM, Lafrankie DC, Ruland S, Wen PY (2011) Retrospective study of dasatinib for recurrent glioblastoma after bevacizumab failure. J Neuro-Oncol 104(1):287–291

20. Bendall SC, Nolan GP, Roederer M, Chatto-padhyay PK (2012) A deep profiler's guide to cytometry. Trends Immunol 33(7):323–332

21. Bandura DR, Baranov VI, Ornatsky OI et al (2009) Mass cytometry: technique for real time single cell multitarget immunoassay based on inductively coupled plasma time-of-flight mass spectrometry. Anal Chem 81(16):6813–6822

22. Bendall SC, Simonds EF, Qiu P et al (2011) Single-cell mass cytometry of differential immune and drug responses across a human hematopoietic continuum. Science 332 (6030):687–696

23. Fellouse F, Esaki K, Birtalan S et al (2007) High-throughput generation of synthetic anti-bodies from highly functional minimalist phage-displayed libraries. J Mol Biol 373 (4):924–940

24. Miersch S, Sidhu SS (2012) Synthetic anti-bodies: concepts, potential and practical consid-erations. Methods 57(4):486–498

25. Amir EAD, Davis KL, Tadmor MD et al (2013) viSNE enables visualization of high dimensional single-cell data and reveals pheno-typic heterogeneity of leukemia. Nat Biotech-nol 31(6):545–552

26. Lathia JD, Gallagher J, Heddleston JM, Wang J, Eyler CE, MacSwords J, Wu Q, Vasanji A, McLendon RE, Hjelmeland AB, Rich JN (2010) Integrin alpha 6 regulates glio-blastoma stem cells. Cell Stem Cell 6 (5):421–432

27. Gangemi RM, Griffero F, Marubbi D, Perera M, Capra MC, Malatesta P, Ravetti GL, Zona GL, Daga A, Corte G (2009) SOX2 silencing in glioblastoma tumor-initiating cells causes stop of proliferation and loss of tumorigenicity. Stem Cells 27(1):40–48

28. Verginelli F, Perin A, Dali R et al (2013) Tran-scription factors FOXG1 and Groucho/TLE promote glioblastoma growth. Nat Commun 4

29. Günther HS, Schmidt NO, Phillips HS et al (2008) Glioblastoma-derived stem cell-enriched cultures form distinct subgroups according to molecular and phenotypic criteria. Oncogene 27(20):2897–2909

30. Qiu P, Simonds EF, Bendall SC et al (2011) Extracting a cellular hierarchy from high-dimensional cytometry data with SPADE. Nat Biotechnol 29(10):886–891

31. Fluidigm Corporation (2014) Maxpar panel designer. https://www.fluidigm.com/binaries/content/documents/fluidigm/search/hippo:resultset/maxpar-panel-designer/fluidigm:file

32. Fluidigm Corporation (2015) Maxpar antibody labeling kit PRD002 Version 8. https://www.fluidigm.com/binaries/content/documents/fluidigm/resources/maxpar-antibody-labeling-kit-pr-prd002/maxpar-antibody-labeling-kit-pr-prd002/fluidigm:file. Accessed 20 May 2017

33. Finck R, Simonds EF, Jager A et al (2013) Normalization of mass cytometry data with bead standards. Cytometry A 83(5):483–494

34. ThermoFisher Scientific (n.d.) Passive adsorp-tion protocol. https://www.thermofisher.com/ca/en/home/life-science/cell-analysis/qdots-microspheres-nanospheres/idc-surfactant-free-latex-beads/latex-bead-protein-coupling-protocols/passive-adsorption-protocol.html

Chapter 15

Pooled Lentiviral CRISPR-Cas9 Screens for Functional Genomics in Mammalian Cells

Michael Aregger, Megha Chandrashekhar, Amy Hin Yan Tong, Katherine Chan, and Jason Moffat

Abstract

CRISPR-Cas9 technology provides a simple way to introduce targeted mutations into mammalian cells to induce loss-of-function phenotypes. The CRISPR-Cas9 system has now successfully been applied for genetic screens in many cell types, providing a powerful tool for functional genomics with manifold applications. Genome-wide guide-RNA (gRNA) libraries allow facile generation of a pool of cells, each harboring a gene knockout mutation that can be used for the study of gene function, pathway analysis or the identification of genes required for cellular fitness. Furthermore, CRISPR genetic screens can be applied for the discovery of genes whose knockout sensitizes cells to drug treatments or mediates drug resistance. Here, we provide a detailed protocol discussing the necessary steps for the successful performance of pooled CRISPR-Cas9 screens.

Key words CRISPR-Cas9, Genetic screen, gRNA library, Knockout, Fitness, Proliferation

1 Introduction

CRISPR-Cas9 technology has revolutionized genetics and offers a simple and inexpensive way for the editing of the human genome such as the generation of mutations resulting in a gene knockout phenotype [1, 2]. When guided to target a locus that shares sequence homology with the guide (g)RNA followed by the proto-spacer adjacent motif NGG, Cas9 induces a double stranded DNA break that, when repaired, often results in insertions or deletions leading effectively to a gene knockout [3]. Pooled gRNA libraries exploit the power of the CRISPR-Cas9 gene editing technology and apply it on a genome-scale level [4]. We and others have developed genome-wide gRNA libraries targeting human protein-coding genes allowing for the simultaneous generation of

Michael Aregger and Megha Chandrashekhar contributed equally to this work.

Sheila K. Singh and Chitra Venugopal (eds.), *Brain Tumor Stem Cells: Methods and Protocols*, Methods in Molecular Biology, vol. 1869, https://doi.org/10.1007/978-1-4939-8805-1_15, © Springer Science+Business Media, LLC, part of Springer Nature 2019

knockouts for nearly every human protein-coding gene [5–13]. The edited cell population can then be exposed to different conditions and the depletion or enrichment of gRNAs over time can be measured by extracting genomic DNA, generating sequencing libraries, and measuring abundance of gRNAs by next generation sequencing. Importantly, CRISPR gene knockout screens outperform shRNA screens, the previous state of the art technology for functional genomics screening in mammalian cells, both in terms of sensitivity and specificity [14].

CRISPR gene knockout libraries can be applied to identify genes whose knockouts cause cellular fitness defects, alter drug sensitivity (sensitizers or resistance genes), regulate the expression of a reporter gene or protein, or are required for a certain cellular state or function of a specific pathway [5–7, 9, 15–19].

Herein, we provide a detailed screening protocol for human cell lines that is optimized for screening with the Toronto Knockout Library version 3.0 (TKOv3) [12]. TKOv3 is a sequence-optimized gRNA library of 71,090 gRNAs targeting 18,053 human protein-coding genes with four gRNAs per gene. In addition, the library contains 142 gRNA sequences targeting EGFP, LacZ, and luciferase for use as controls in experiments using these reporter genes. TKOv3 is a one-component lentiviral library containing both gRNA and Cas9 expression cassettes allowing the simultaneous delivery of the components required for gene editing to any cell type amenable to lentiviral transduction. Like its precursor TKOv1 [5], which is a two-component lentiviral library that requires the prior generation of a Cas9 expressing cell line, TKOv3 can be obtained through Addgene (www.addgene.org). This protocol describes the required steps to successfully perform a pooled CRISPR gene knockout screen including amplification of the library, production of library lentivirus, performing the screen, extracting genomic DNA, generating sequencing libraries, sequencing of experimental samples and brief highlights on current data analysis methods.

2 Materials

2.1 Tissue Culture

1. The entire assay procedure has to be performed in a Class IIa biosafety cabinet and 37 °C, 5% CO_2 cell incubator.

2. Media requirements (culture medium, serum, growth factors, etc.) and conditions for adherence to tissue culture vessels for the desired cell lines have to be met.

3. Typically, the cell culture for pooled screens is done in 15 cm plates or T160 flasks.

2.2 Library Amplification

1. CRISPR gRNA library, e.g., TKOv3 (Addgene: Pooled Library #90294).

2. Electroporator, electroporation cuvettes.

3. Electrocompetent cells (e.g., Lucigen Endura electrocompetent cells).

4. SOC recovery medium, LB agar plates with carbenicillin, LB medium with carbenicillin.

5. 37 °C shaking incubator and 30 °C plate incubator.

6. Plasmid Maxi or Mega Purification Kit.

2.3 Virus Production

1. 293T packaging cells (recommended: passage number < 15).

2. Transfection quality plasmids: CRISPR gRNA library, psPAX2 (packaging plasmid), pMD2.G (envelope plasmid).

3. X-tremeGENE 9 DNA Transfection reagent (Roche).

4. Opti-MEM serum-free media.

5. Cell seeding media — Low-antibiotic growth media (DMEM + 10% FBS + 0.1× Pen/Strep): 500 mL DMEM (Dulbecco's Modification of Eagle's Medium) + 50 mL FBS (heat-inactivation of FBS is optional) + 0.5 mL 100× Pen/Strep.

6. Viral harvest media — Serum-free, high-BSA 293T growth media (DMEM + 1.1 g/100 mL BSA + 1× Pen/Strep): 500 mL DMEM + 32 mL 20 g/100 mL BSA stock (dissolved in DMEM, filter sterilized) + 5 mL 100× Pen/Strep.

2.4 Pooled CRISPR Knockout Screen

1. Polybrene (2 mg/mL).

2. Puromycin (2 mg/mL).

3. QIAamp Blood Maxi Kit.

4. RNase A (100 mg/mL).

5. Centrifuge with swinging-bucket rotor.

6. 5 M NaCl.

7. NEBNext Ultra II Q5 Master Mix.

8. PCR primers as listed in Tables 1 and 2.

9. Blue-light Transilluminator.

10. Nanodrop or Qubit.

11. DNA Gel extraction kit.

12. Low molecular weight DNA ladder.

Table 1
PCR Primers for TKOv3 sequencing libraries (*see* Note 27)

PCR 1 primers are regular desalted oligos.

PCR 2 primers can be ordered from IDT as desalted Ultramer DNA oligos.

PCR 1 Primer sequences

v2.1-F1

GAGGGCCTATTTCCCATGATTC

v2.1-R1

GTTGCGAAAAAGAACGTTCACGG

PCR 2 Primer sequences—i5 forward primers

D501-F

AATGATACGGCGACCACCGAGATCTACACTATAGCCTACACTCTTTCCCTACACGACGCT
CTTCCGATCT**TTGTGGAAAGGACGAAACACCG**

D502-F

AATGATACGGCGACCACCGAGATCTACACATAGAGGCACACTCTTTCCCTACACGACGCT
CTTCCGATCT**TTGTGGAAAGGACGAAACACCG**

D503-F

AATGATACGGCGACCACCGAGATCTACACCCTATCCTACACTCTTTCCCTACACGACGC
TCTTCCGATCTTTGTGGAAAGGACGAAACACCG

D504-F

AATGATACGGCGACCACCGAGATCTACACGGCTCTGAACACTCTTTCCCTACACGACGC
TCTTCCGATCT**TTGTGGAAAGGACGAAACACCG**

D505-F

AATGATACGGCGACCACCGAGATCTACACAGGCGAAGACACTCTTTCCCTACACGACGC
TCTTCCGATCT**TTGTGGAAAGGACGAAACACCG**

D506-F

AATGATACGGCGACCACCGAGATCTACACTAATCTTAACACTCTTTCCCTACACGACGCTC
TTCCGATCT**TTGTGGAAAGGACGAAACACCG**

PCR 2 Primer sequences—i7 reverse primers

D701-R

CAAGCAGAAGACGGCATACGAGATCGAGTAATGTGACTGGAGTTCAGACGTGTGCTCTT
CCGATCT**ACTTGCTATTTCTAGCTCTAAAAC**

D702-R

CAAGCAGAAGACGGCATACGAGATTCTCCGGAGTGACTGGAGTTCAGACGTGTGCTC
TTCCGATCT**ACTTGCTATTTCTAGCTCTAAAAC**

D704-R

CAAGCAGAAGACGGCATACGAGATGGAATCTCGTGACTGGAGTTCAGACGTGTGCTCTT
CCGATCT**ACTTGCTATTTCTAGCTCTAAAAC**

(continued)

Table 1
(continued)

D705-R

CAAGCAGAAGACGGCATACGAGAT<u>TTCTGAAT</u>GTGACTGGAGTTCAGACGTGTGCTCTT
CCGATCT**ACTTGCTATTTCTAGCTCTAAAAC**

D706-R

CAAGCAGAAGACGGCATACGAGAT<u>ACGAATTC</u>GTGACTGGAGTTCAGACGTGTGCTCTT
CCGATCT**ACTTGCTATTTCTAGCTCTAAAAC**

D707-R

CAAGCAGAAGACGGCATACGAGAT<u>AGCTTCAG</u>GTGACTGGAGTTCAGACGTGTGCTCTTC
CGATCT**ACTTGCTATTTCTAGCTCTAAAAC**

<u>Underlined sequence</u> denotes i5 or i7 index (*see* Table 2)
Bold sequence denotes annealing sequence

Table 2
Indexes for multiplexed Illumina sequencing

i7 Index name	i7 Sequence	i7 Index for demultiplexing of sequencing reads	i5 Index name	i5 Sequence	i5 Index for demultiplexing of sequencing reads for HiSeq	i5 Index for demultiplexing of sequencing reads for NextSeq
D701	CGAGTAAT	ATTACTCG	D501	TATAGCCT	TATAGCCT	AGGCTATA
D702	TCTCCGGA	TCCGGAGA	D502	ATAGAGGC	ATAGAGGC	GCCTCTAT
D704	GGAATCTC	GAGATTCC	D503	CCTATCCT	CCTATCCT	AGGATAGG
D705	TTCTGAAT	ATTCAGAA	D504	GGCTCTGA	GGCTCTGA	TCAGAGCC
D706	ACGAATTC	GAATTCGT	D505	AGGCGAAG	AGGCGAAG	CTTCGCCT
D707	AGCTTCAG	CTGAAGCT	D506	TAATCTTA	TAATCTTA	TAAGATTA

3   Methods

3.1   gRNA Library Amplification

1. Dilute the library to 50 ng/μL in Buffer TE (*see* **Note 1**).

2. Electroporate the library. Set up 4 electroporations as follows: add 2 μL of 50 ng/μL CRISPR gRNA library to 25 μL of Lucigen Endura electro-competent cells. Electroporate according to the manufacturer's suggested conditions and protocol. Add 975 μL of Recovery Medium (or SOC medium) and transfer cells to a culture tube with an additional 1 mL of Recovery Medium. Place tubes in a shaking incubator at 200 rpm for 1 h at 37 °C.

3. Titer the library (*see* **Note 2**). Pool all 8 mL of recovered cells. Mix well. Transfer 10 μL of cells to 990 μL of Recovery Medium, mix well, and plate 20 μL onto a pre-warmed

10-cm LB + carbenicillin agar plate (this is a 40,000-fold dilution of the full transformation).

4. Plate the library. Spread recovered cells on a total of 20 15-cm LB + carbenicillin agar plates. Spread 400 µL of recovered cells on each of the pre-warmed plates.

5. Incubate the plates for 14–16 h at 30 °C.

6. Calculate transformation efficiency. Count the number of colonies on the dilution plate. Multiply this number of colonies by 40,000 for the total number of colonies plated. Proceed if the total number of colonies represents a library coverage of at least 200-fold (optimally 500–1000). Otherwise optimize the transformation protocol or increase the number of electroporations (*see* **Note 3**).

7. Harvest colonies. Transfer 7 mL of LB + carbenicillin medium on one 15 cm plate. Scrape the colonies off with a cell spreader and transfer the scraped cells into a sterile bottle using a 10 mL pipet. Rinse the scraped plate with an additional 5 mL of LB + carbenicillin medium and transfer to the bottle.

8. Pool all scraped cells from 20 plates to the bottle. Stir well for 1 h at room temperature.

9. Transfer cells to pre-weighed centrifuge bottles. Centrifuge at $7000 \times g$ and discard media. Invert the bottles to allow draining of remaining supernatant. Weigh the wet cell pellet.

10. Purify the library plasmid pool using a maxi- or mega-scale plasmid purification kit. Perform multiple maxi or mega preps according to column capacity. Typically, a maxi column can process 1 g of wet cell pellet, and a mega column can process 2.5 g of wet cell pellet.

3.2 Large-Scale CRISPR gRNA Lentivirus Production

1. Seed HEK293T packaging cells in low-antibiotic growth media at 8E6–9E6 cells per 15 cm plate in 20 mL media. For 1 L virus production prepare ~60 plates (*see* **Note 4**).

2. Incubate cells for 24 h (37 °C, 5% CO_2), or until the following afternoon. The cells should be 70% −80% confluent at moment of transfection.

3. Transfect HEK293T packaging cells. Prepare a mixture of the three transfection plasmids (~1:1:1 molar ratio) in Opti-MEM and a separate mixture of Opti-MEM and X-tremeGENE as outlined below (for TKOv1 *see* **Note 5**). Following 5 min incubation, add appropriate amount of plasmid mix to X-tremeGENE mix dropwise for a 3:1 ratio of transfection reagent:DNA complex. Mix by gently flicking the tube. Incubate the transfection mix for 30 min. Carefully transfer the transfection mix to the packaging cells by adding dropwise in circular, zigzag motion.

	Reagents per 15 cm plate
Opti-MEM	100 µL
psPAX2	4.8 µg
pMD2.G	3.2 µg
TKOv3 CRISPR gRNA library	8 µg

	Reagents per 15 cm plate
Opti-MEM	800 µL
X-tremeGENE	48 µL

4. Incubate cells for 18 h (37 °C, 5% CO_2).

5. Change media to remove the transfection reagent and replace with high-BSA growth media for viral harvests (18–20 mL/ 15 cm plate).

6. Incubate cells for 24 h (37 °C, 5% CO_2).

7. Next day, harvest media containing lentivirus at ~40 h post-transfection. Transfer media to a polypropylene storage tube (*see* **Note 6**).

8. Spin the media containing virus at 500 × *g* for 5 min to pellet any packaging cells that were collected during harvesting. Aliquot supernatant into sterile screw cap polypropylene storage tube (*see* **Note 7**).

3.3 Cell Line Characterization

1. Choose desired cell line to be screened and ensure cell line is mycoplasma-free before starting and after completing the screen.

2. Determine optimal cell plating density. Typically, the cell culture for pooled screens is done in 15 cm plates and cells should be passaged before reaching full confluency. Measure the approximate doubling time of your cells under the chosen condition. Ensure that cells adhere reasonably well to tissue culture vessels.

3. Determine puromycin sensitivity of cell line by doing a kill curve. Dilution range should span 0 µg/mL to 5 µg/mL in 0.5 µg/mL increments. Concentration of puromycin to be used in a screen should be 0.5–1 µg/mL higher than that required to kill 100% of uninfected cells in 48 h.

4. Test cells for sensitivity to either polybrene (up to 8 µg/mL) or protamine sulfate (up to 5 µg/mL) by doing a dose response curve in the same method as used for measuring puromycin sensitivity.

5. Perform "screenability assay" to determine editing efficiency of the chosen cell line. For this assay, infect cells with lentiviral gRNA vectors targeting noncoding controls such as *AAVS1*

(negative control) and exonic sequences of essential genes such as *PSMD1* and *PSMB2* (positive controls) respectively. After 48 h of puromycin selection set up a proliferation assay and determine cell viability 7–10 days post-seeding by cell counting or viability dye. A strong proliferation defect should be visible upon targeting essential genes in cell lines with good editing efficiency (*see* **Notes 8** and **9**).

6. If a two-component gRNA library is chosen (i.e., library plasmid does not contain Cas9 expression cassette; e.g., TKOv1), cells need to be engineered to express Cas9. For the generation of Cas9 expressing cells please refer to Hart et al. Cell [5]. Before starting a screen it is crucial to determine editing efficiency by the screenability assay described above. To increase the performance of the screen a single cell clone expressing Cas9 with high editing efficiency may be selected (*see* **Note 10**).

3.4 Lentivirus Multiplicity of Infection (MOI) Determination

1. Thaw a fresh aliquot of pooled CRISPR gRNA library lentivirus (keep on ice).

2. Design dilution series of virus for MOI determination between 0 and 2 mL.

3. Plate cells, add media and polybrene (typically final concentration of 8 μg/mL), and then add your designated volume of virus. Mix plates thoroughly by tilting for 2 min before transferring them to incubator (*see* **Notes 11** and **12**).

4. Add the same volume of virus to two vessels for each dilution point.

5. 24 h after addition of virus, cells should be infected and tightly adhered to plate. Remove media using pipettes (dispose of media and pipettes in 20% bleach solution).

6. Gently wash plates with PBS to remove any extraneous virus (dispose of PBS and pipettes in 20% bleach solution).

7. Add fresh media (20 mL for 15 cm plate) containing puromycin at the required concentration to vessels of one virus dilution series and fresh media without puromycin to the other virus dilution series.

8. After 48 h of puromycin selection, all uninfected cells should be dead. You must use a dose of puromycin that will kill all uninfected cells within 48 h.

9. Remove media, wash cells with PBS to dislodge remaining dead cells, and add trypsin to collect selected cells.

10. Count cells from all plates and graph results for the two series (+/− puromycin).

11. Determine virus volume that gives 30–40% survival with puromycin selection vs. without puromycin. This is the volume of pooled virus that gives an MOI of 0.3–0.4 under the tissue culture conditions used.

3.5 Pooled CRISPR Knockout Screen

3.5.1 Primary Screen Infection and Cell Passaging

1. Expand cells to approximately 8E7–9E7 cells for Day 1.

2. The total number of cells plated for infection should be such that with infection at MOI 0.3, puromycin selection and the growth rate of the cells, ~1.20E8 cells can be harvested at T0 (end of puromycin selection).

3. Infections are done in 15 cm TC plates at 200-fold coverage of the CRISPR gRNA library (*see* **Note 13**). Extra plates are prepared to accommodate MOI fluctuations and control plates are also required. The following plates are needed:

	Treatment
Screening plates	Virus, + puromycin
Control 1	No virus, + puromyocin (0% survival control)
Control 2	Virus, no puromycin (100% survival control)

4. Harvest cells and pool into a sterile vessel. Seed required cell number to each plate.

5. Add virus and polybrene (final concentration 8 μg/mL) to plates (*see* **Note 14**).

6. Mix plates thoroughly by titling the plates for 2 min.

7. 24 h after addition of virus, cells should be infected and tightly adhered to plate.

8. Remove media using pipette (dispose of media and pipettes in 20% bleach solution) and gently wash plate with PBS to remove any extraneous virus (dispose of PBS and pipettes in 20% bleach solution).

9. Add fresh media containing puromycin at the required concentration to the cells.

10. 48 h after puromycin addition, all uninfected cells should be dead (control 1).

11. Remove media, gently wash plate with PBS to dislodge remaining dead cells (*see* **Note 15**).

12. Trypsinize and collect cells from all plates into one sterile container. Make sure that all cell clumps are dispersed by gentle repeated pipetting when harvesting cells from plates.

13. Count cells and calculate number of cells/mL.

14. Collect three replicates at 200-fold library coverage each by centrifugation (*see* **Note 16**): Spin at 335 × *g* for 5 min. Wash

with PBS. Label tubes and freeze the cell pellets "dry" at −80 °C.

15. From the pool, plate cells into three replicate groups. For dropout screens, 200-fold coverage of the library should be maintained throughout the screen for each replicate, while positive selections screens can be performed at 50–100-fold representation (*see* **Note 17**). For positive selection screens refer to comment in **Note 17** (*see* **Note 18**).

16. Passage the cells every 3–8 days out to the end-point (~3 weeks) following the steps outlined below (*see* **Note 19**).

17. Trypsinize and collect cells. Pool cells from all the vessels in each separate replicate group with each other (*see* **Note 20**).

18. Re-plate cells at 200-fold coverage (i.e., ~1.5E7 for TKOv3) for each replicate group exactly as done at T0.

19. Collect and freeze cell pellets at 200-fold library representation for each replicate group (i.e., ~1.5E7 for TKOv3). Each pellet gets a time (T) number and a replicate designation. This number corresponds to the number of days post T0 that it was collected (e.g., T3_A, T6_B, T9_C, etc.) (*see* **Note 21**).

3.5.2 Genomic (g)DNA Extraction and Precipitation

1. gDNA extraction is performed using the QIAamp Blood Maxi kit (*see* **Note 22**) essentially as described in the kit manual with the addition of RNase A treatment and modification of elution volume. Follow the instructions below, using the indicated volumes and refer to the kit manual for extra details if required (*see* **Notes 1** and **23**). It is very important to use a swinging-bucket rotor in a centrifuge that is capable of attaining the required g-force. Failure to do so will result in low yield and dirty genomic DNA.

2. Prepare 70 °C water/bead bath with a rack for your tubes. Ensure that buffers are prepared and ready to use. Prepare Qiagen protease solution in water as instructed on the bottle.

   Warm up buffer AE to 65 °C (aliquot needed amount in 50 mL tube).

3. Thaw cell pellets at room temperature for 5–10 min. Label samples and prelabel tubes and columns to be used in extraction.

4. Add 4.5 mL of sterile PBS to each cell pellet. Resuspend cells with a P1000 pipette. Seal the tube tightly and vortex thoroughly to disperse cells. Pipette to disrupt cell clumps if needed.

5. Add 105 μL of RNase A (100 mg/mL) to resuspended cells. Mix briefly by swirling. Incubate at room temperature for 5 min.

6. Add 500 μL of Qiagen protease solution to each sample. Mix briefly by swirling.

7. Add 6 mL Buffer AL. Cap tube and mix by inversion for 2 min.

8. Incubate the tube in 70 °C water bath for 15 min.

9. Let the tube cool to ~40 °C (leave at room temperature for 30 min).

10. Add 5 mL ethanol (96–100%). Mix by shaking and inverting for 2 min.

11. Using a 10 mL pipette, carefully transfer the solution into a prelabeled QIAamp maxi column placed in the provided 50 mL centrifuge tube. Take care not to spill onto the rim of the column.

12. Centrifuge the tube at 1850 × g for 3 min at room temperature.

13. Transfer column into a new, appropriately labeled 50 mL centrifuge tube.

14. Using a 10 mL pipette, carefully transfer the flow-through and reapply to the column for a second binding step. Take care not to spill onto the rim of the column.

15. Centrifuge the tube again at 1850 × g for 3 min at room temperature.

16. Transfer the QIAamp maxi column to a new 50 mL centrifuge tube.

17. Add 5 mL Buffer AW1 to the column, being careful to not spill onto the rim.

18. Cap the tube and centrifuge at 4500 × g for 2 min at room temperature.

19. Add 5 mL Buffer AW2 to the column, being careful to not spill onto the rim.

20. Cap the tube and centrifuge at 4500 × g for 15 min at room temperature.

21. Check the tubes—if any buffer remains on the inside edge of the column, or if the filter appears to be wet, uncap the columns and dry in a warm incubator for ~15 min.

22. Place the QIAamp maxi column into clean a 50 mL centrifuge tube, and discard the filtrate and the previous tubes.

23. Pipet 1 mL of Buffer AE (pre-warmed 65 °C) directly onto the membrane and cap the tube.

24. Incubate the tube at room temperature for 10 min, then centrifuge at 4500 × g for 15 min.

25. Pipet additional 1 mL of Buffer AE (pre-warmed 65 °C) directly onto the membrane and cap the tube.

26. Incubate the tube at room temperature for 10 min, then centrifuge at 4500 × *g* for 15 min.

27. Collect flow through and transfer to a 1.5 mL microfuge tube. Vortex to mix sample thoroughly.

28. Quantitate DNA and measure purity by spectrophotometry, then freeze at −20 °C or proceed with gDNA precipitation.

29. Vortex each sample thoroughly to mix the genomic DNA.

30. Split samples into 1.5 mL microfuge tubes containing 500 μL each (four tubes per sample).

31. Add 20 μL of 5 M NaCl to a final concentration of 0.2 M and 2 volumes (1000 μL) of 96–100% ethanol. Do not use sodium acetate as residual acetate interferes with downstream PCR reaction.

32. Invert tubes 10 times (until thoroughly mixed), then spin at 16,500 × *g* for 15 min at 4 °C in a table-top centrifuge.

33. Aspirate the supernatant, being careful to avoid DNA pellet.

34. Add 500 μL 70% ethanol. Wash pellet by inverting tube 5 times, then centrifuge at 16,500 × *g* for 15 min at 4 °C.

35. Aspirate the supernatant. Pulse-spin down any remaining liquid and aspirate it.

36. Air dry pellet ~10 min (not longer unless pellet is still wet) and re-solubilize each replicate in 75 μL of buffer TE (300 μL total volume).

37. Heat samples at 50 °C for 1 h (or until completely solubilized). Gently vortex at 20 min intervals to fully resuspend/solubilize the DNA (*see* **Note 24**).

38. Pool samples and spin at 16,500 × *g* for 10 min to remove remaining clumps.

39. Transfer supernatant to low DNA binding tubes and determine DNA concentration with Nanodrop or Qubit using Qubit dsDNA BR Assay (recommended).

3.5.3 CRISPR Sequencing Library Preparation

1. The following sequencing library protocol is optimized for TKOv3 (modified lentiCRISPRv2 backbone). For TKOv1 primers and protocol please refer to Hart et al. [5].

2. Perform two PCR reactions to (1) enrich guide-RNA regions in the genome and (2) amplify guide-RNA with Illumina TruSeq adapters with i5 and i7 indices (*see* **Notes 1** and **25–27** and Tables 1 and 2).

3. Set up PCR 1 using total of 50 μg of genomic DNA. Add 2.5 μg of genomic DNA per 50 μL reaction. Set up 20 identical 50 μL reactions to achieve the desired coverage (*see* **Note 28**).

Always perform one PCR without any genomic DNA template in order to ensure there is no contaminating template (for both PCR steps).

	1×
2× NEBNext Ultra II Q5 Master Mix	25 µL
10 µM v2.1-F1	2.5 µL
10 µM v2.1-R1	2.5 µL
Genomic DNA	2.5 µg
Water	up to 50 µL
Total	50 µL

4. Amplify reactions in a thermocycler using the following program:

Step	Temperature	Time	
1	98 °C	30 s	
2	98 °C	10 s	25 cycles (**steps 2–4**)
3	66 °C	30 s	
4	72 °C	15 s	
5	72 °C	2 min	
6	10 °C	Hold	

5. Run 2 µL of PCR 1 product on a 1% agarose gel. Visualize the PCR product on a gel imager. PCR 1 yields a product of 600 bp (*see* **Note 29**).

6. Pool all individual 50 µL reactions for each genomic DNA sample, mix by vortex.

7. Set up PCR 2 using unique i5 and i7 index primer combinations for each individual sample to allow pooling of sequencing libraries. Set up one 50 µL reaction for each sample. Use 5 µL of the pooled PCR 1 product as template.

	1×
2× NEBNext Ultra II Q5 Master Mix	25 µL
10 µM i5 primer	2.5 µL
10 µM i7 primer	2.5 µL
PCR 1 product	5 µL
Water	15 µL
Total	50 µL

8. Amplify reaction in a thermocycler using the following program:

Step	Temperature	Time	
1	98 °C	30 s	
2	98 °C	10 s	10 cycles (**steps 2–4**)
3	55 °C	30 s	
4	65 °C	15 s	
5	65 °C	5 min	
6	10 °C	Hold	

9. Run 50 μL of PCR 2 product on a 2% agarose gel. PCR 2 yields a product of 200 bp (*see* **Note 30**).

10. Visualize the PCR product on a Dark Reader (blue light transilluminator, not UV-light). Excise the 200 bp band and purify DNA from agarose gel slice using a gel extraction kit.

11. Quantitate and measure purity of sequencing library on both NanoDrop and Qubit.

3.5.4 High-Throughput Sequencing

1. The resulting libraries can be pooled and sequenced on Illumina HiSeq 2500 or NextSeq 500 with standard primers for dual indexing. For HiSeq 2500, use standard Single-Read (SR) 50-cycle chemistry with dual-index. For NextSeq 500, use standard Single-Read (SR) 75-cycle chemistry with dual-index.

2. The following sequencing strategy is recommended: Dark Cycle: 21 bp; Read 1: 26 bp; Index Read 1: 8 bp; Index Read 2: 8 bp (*see* **Notes 31** and **32**).

3. For dropout screens it is recommended to read each sample at 200-fold library coverage. A higher coverage of 500-fold is recommended for T0 samples. For strong positive selection screens a 50-fold coverage may be sufficient for the identification of enriched gRNAs.

4. Always include T0 sample to determine library representation for the particular screen and for the determination of fold change of gRNAs over time.

3.5.5 Data Analysis

1. Demultiplex sequencing data and map reads to the gRNA sequence library using Bowtie with –v2 (allowing two mismatches) and –m1 (discarding any read that mapped to more than one sequence in the library) (see Table 2).

2. Normalize read counts to ten million reads per sample.

3. Calculate the fold change of each gRNA for each replicate of each time point versus the T0 sample. gRNAs with less than

30 raw (non-normalized) reads in the T0 sample should be excluded for fold-change calculation and downstream analysis.

4. Assess screen performance by determining fold change of reference essential and nonessential genes [12, 20, 21].

5. Analyze fold-changes with the BAGEL algorithm to calculate a Bayes Factor for each gene representing a confidence measure that the gene knockout results in a fitness defect (*see* **Note 33**) [5, 12, 21].

4 Notes

1. For all steps of this protocol it is crucial to work in a clean environment to avoid cross contamination with other plasmid DNA or gDNA, which can substantially interfere with results. Cleaning the working surface as well as pipettes and racks with 20% bleach is recommended. Carefully wipe all bleached material with water to remove bleach. Aliquot out all required reagents to reduce contamination. Use filter tips for all steps and avoid creating aerosols when ejecting tips.

2. The dilution plate is used to estimate transformation efficiency to ensure that full library representation is preserved.

3. For example, for the TKOv3 library the minimal colony number is at least 1.4×10^7. This efficiency is equivalent to $200\times$ colonies per guide-RNA construct in the 71,090 gRNA TKOv3 library. For the TKOv1 library (~90,000 gRNAs) the minimal colony number would be 1.8×10^7.

4. Ensure that all cell lines are of mycoplasma-free before virus production and starting screens. There are many commercial mycoplasma detection kits available.

5. Plasmid ratio for TKOv1 (pLCKO vector): 5.8 µg of psPAX2 and 4.2 µg of pMDG.2, 6 µg of TKOV.1 library plasmid (1:1:1 molar ratio) in 80 µL of Opti-MEM.

6. Lentiviral precautions: When working with active lentivirus (production, MOI determination and primary infection steps), double gloves must be worn and no waste can exit the hood until treated with bleach. No vacuum pumps can be used, liquid waste must be manually removed and bleached for 30 min before discarding.

7. Virus may be stored at 4 °C for short periods (hours to days), but should be frozen at −80 °C for long-term storage. To reduce the number of freeze/thaw cycles, aliquot large-scale virus preps to smaller storage tubes prior to long-term storage.

8. Use the same vector format as the chosen gRNA library (e.g., lentiCRISPRv2 one-component vector for TKOv3). Seed cells

at low density to allow proliferation over a period of 7–10 days. Do not passage cells during the experiments to avoid selecting for cells with in-frame editing.

9. In addition to the proliferation-based assay surveyor assay or western blotting can be performed to determine editing on DNA or protein level respectively.

10. Working with a pooled population of Cas9 expressing cells (i.e., Cas9 cell line or Cas9 delivered via one-component library) works well for most cell lines. Selecting a single cell clone with high editing efficiency may increase the performance of the screen but may introduce a clonal effect.

11. The multiplicity of infection (MOI) must be determined under the same cell culture conditions used during the primary screen. This includes using the same tissue culture vessels, media constituents and volume, cell plating density, and pooled virus preps (without prior thaws) that will be used in the screen. Measurements made in different formats (e.g., 6-well plates) cannot be reliably scaled to the screening format.

12. For the determination of the MOI it is important that the non-selected cells keep proliferating during the assay to avoid that the selected cells can catch up due to lower cell confluence. Thus, a seeding density needs to be chosen that does not result in reaching full confluency and growth arrest within 72 h.

13. For a 200-fold coverage of the TKOv3 library at an MOI 0.3, 5E7 cells are required for infection at 3E6–5E6 cells/plate. For the TKOv1 library, 6E7 cells are required.

14. Alternatively, batch infections can be done by adding virus and polybrene to cells in suspension, mixing well, then plating.

15. It is important to remove all dead cells since their genomic DNA can contaminate the T0 prep. Wash twice with PBS if necessary.

16. For TKOv3 collect 2E7 cells to reach about 250-fold coverage of the library. For TKOv1 collect 2.25E7 cells. These are the time zero (T0) samples. Two of the three replicates will serve as backup samples and may not be processed further. Collecting more cells than the required 200-fold will provide a buffer for potential material loss during genomic DNA extraction.

17. 200-fold library representation should be considered a minimum for dropout screens (e.g., identification of essential genes, sensitizers to drugs, etc.). If desired, a higher representation can be maintained. For TKOv3 (71,090 gRNAs), each replicate should contain in total 1.5E7 cells across the required number of vessels (use exactly same number of cells for each replicate plate, and exactly same number of total cells between each replicate- example Replicate A—seed 3 plates at

5E6/plate for total 1.5E7 cells, repeat for Replicate B and C). Do not use puromycin in this or in subsequent plating steps as this may affect proliferation rates. For TKOv1 plate 1.8E7 cells per replicate.

18. For positive selection screens (e.g., drug resistance screen) in which only a small cell population is expected to survive a specific culture condition the library coverage can be reduced to 50–100-fold.

19. Most adherent cell lines require splitting every 3–8 days. The pool-infected cells should be passaged at the same density that cells would normally be split when expanding them. A transient slowing of cell growth after the infection may be noticed, with a return to normal growth speed 5–15 days after infection. The length and severity of this effect is cell-line dependent.

20. For example, all the cells from replicate A are remixed together from separate plates, all the cells from replicate B are remixed together from separate plates, etc.). This serves to minimize stochastic growth effects in random subpopulations in individual vessels within a replicate group.

21. Note: pelleting more cells (i.e., 2E7) than the 1.5E7 required for maintaining 200-fold representation helps to buffer for material loss during downstream genomic DNA processing procedures. If you chose to pellet more cells, make sure that all pellets collected in the screen have the same number of cells.

22. As a cost- and time-effective alternative to the QIAamp Blood Maxi kit, we can also recommend the Wizard Genomic DNA Purification Kit (Promega) according to the manufacturer's instructions.

23. This protocol is optimized for cell pellets containing 2E7–5E7 cells (we generally use 2E7–3E7 cells).

24. Make sure that DNA is fully solubilized with no lumps or clumps.

25. Setting up PCR reactions in a dedicated PCR hood is recommended. Treating the agarose gel equipment with 0.1 N HCl for 30 min prior to casting gels is recommended to reduce contamination with plasmids or other sequencing libraries.

26. As an alternative to the two-step PCR approach sequencing libraries can be generated in a single PCR protocol similar to the one described in Hart et al., Cell [5]. For a one-step PCR approach use the barcoded PCR2 primers as described herein (note that larger volumes of the barcoded primers are required for the one-step approach).

27. If there is no option to run dark cycles, i5 forward primers have to be modified to include stagger regions of different length. These stagger regions will maintain the sequence diversity

across the flow-cell at each imaging cycle by shifting apart the constant sequences between each samples. Staggered primers can also be used in combination with dark cycles to increase sequence diversity when reading beyond the guide sequence reaching the constant tracrRNA sequence.

Stagger regions need to be introduced right before the annealing sequence of the i5 forward primer as outlined below, and we recommend varying the length of the stagger from 0 to 5 bases (N refers to the i5 barcode, S refers to the staggered region):

i5 forward primer:
AATGATACGGCGACCACCGAGATCTACAC[NNNNNN NN]ACACTCTTTCCCTACACGACGCTCTTCCGATCT [S]TTGTGGAAAGGACGAAACACCG

The i7 reverse primers do not need to be modified.

Importantly, when introducing stagger regions the sequencing strategy needs to be adjusted by increasing the Read 1 as follows: 26 bp + [length of longest stagger region].

In addition to adding stagger regions, the sequence diversity across the flow-cell can be further increased by using primer pairs in reversed orientation (i.e., the i5 forward primer binds to the tracrRNA region and the i7 reverse primer binds to the U6 promoter).

28. Assuming a diploid human genome is ~6.5 μg and one guide-RNA per genome, 50 μg of genomic DNA yields ~100-fold coverage of the TKO v3 library or 85-fold coverage of the TKOv1 library.

29. A faint band above the major 600 bp band may be visible.

30. Multiple bands may be visible in the product of PCR 2 with a major band at 200 bp. Thus, it is important to separate the bands well at low voltage over a 1–1.5 h run.

31. 20 dark cycles, or base additions without imaging, are applied to get through the constant region covering the end of the U6 promoter. The actual read begins after the dark cycles.

32. When using staggered primers the sequencing strategy needs to be adjusted by increasing the Read 1 as follows: 26 bp + [length of longest stagger region].

33. Use the updated version of the core essential gene reference set (CEG2) for optimal analysis [12]. Note the BAGEL algorithm works best for the detection of gRNAs that drop out from a given population [5, 21]. In addition to BAGEL, there are a number of other algorithms designed for the identification of both enriched and depleted gRNAs. CRISPR-AnalyzeR.org offers a web-based analysis platform for pooled CRISPR screens.

Acknowledgments

We would like to thank members of the Moffat lab for helpful comments. This work was supported by the Canadian Institutes for Health Research (CIHR#342551) to JM. MA holds a postdoctoral fellowship award from the Swiss National Science Foundation, MC holds an Ontario Graduate Scholarship and JM is a Canada Research Chair in Functional Genetics Tier 2. Michael Aregger and Megha Chandrashekhar contributed equally to this work.

References

1. Barrangou R, Doudna JA (2016) Applications of CRISPR technologies in research and beyond. Nat Biotechnol 34:933–941. https://doi.org/10.1038/nbt.3659

2. Wright AV, Nuñez JK, Doudna JA (2016) Biology and applications of CRISPR systems: harnessing nature's toolbox for genome engineering. Cell 164:29–44. https://doi.org/10.1016/j.cell.2015.12.035

3. Jiang F, Doudna JA (2017) CRISPR–Cas9 structures and mechanisms. Annu Rev Biophys 46. https://doi.org/10.1146/annurev-biophys-062215-010822

4. Joung J, Konermann S, Gootenberg JS et al (2017) Genome-scale CRISPR-Cas9 knockout and transcriptional activation screening. Nat Protoc 12:828–863. https://doi.org/10.1038/nprot.2017.016

5. Hart T, Chandrashekhar M, Aregger M et al (2015) High-resolution CRISPR screens reveal fitness genes and genotype-specific cancer liabilities. Cell 163:1515–1526. https://doi.org/10.1016/j.cell.2015.11.015

6. Wang T, Wei JJ, Sabatini DM, Lander ES (2014) Genetic screens in human cells using the CRISPR-Cas9 system. Science 343:80–84. https://doi.org/10.1126/science.1246981

7. Wang T, Birsoy K, Hughes NW et al (2015) Identification and characterization of essential genes in the human genome. Science 350:1096–1101. https://doi.org/10.1126/science.aac7041

8. Park RJ, Wang T, Koundakjian D et al (2016) A genome-wide CRISPR screen identifies a restricted set of HIV host dependency factors. Nat Genet 49:193–203. https://doi.org/10.1038/ng.3741

9. Tzelepis K, Koike-Yusa H, De Braekeleer E et al (2016) A CRISPR dropout screen identifies genetic vulnerabilities and therapeutic targets in acute myeloid leukemia. Cell Rep 17:1193–1205. https://doi.org/10.1016/j.celrep.2016.09.079

10. Ma H, Dang Y, Wu Y et al (2015) A CRISPR-based screen identifies genes essential for west-nile-virus-induced cell death. Cell Rep 12:673–683. https://doi.org/10.1016/j.celrep.2015.06.049

11. Doench JG, Fusi N, Sullender M et al (2016) Optimized sgRNA design to maximize activity and minimize off-target effects of CRISPR-Cas9. Nat Biotechnol 34:184–191. https://doi.org/10.1038/nbt.3437

12. Hart T, Tong A, Chan K et al (2017) Evaluation and design of genome-wide CRISPR/SpCas9 Knockout screens. G3 7(8):2719–2727. https://doi.org/10.1534/g3.117.041277

13. Sanjana NE, Shalem O, Zhang F (2014) Improved vectors and genome-wide libraries for CRISPR screening. Nat Methods 11:783–784. https://doi.org/10.1038/nmeth.3047

14. Evers B, Jastrzebski K, Heijmans JPM et al (2016) CRISPR knockout screening outperforms shRNA and CRISPRi in identifying essential genes. Nat Biotechnol 34:631–633. https://doi.org/10.1038/nbt.3536

15. Shalem O, Sanjana NE, Hartenian E et al (2014) Genome-scale CRISPR-Cas9 knockout screening in human cells. Science 343:84–87. https://doi.org/10.1126/science.1247005

16. Koike-Yusa H, Li Y, Tan E-P et al (2013) Genome-wide recessive genetic screening in mammalian cells with a lentiviral CRISPR-guide RNA library. Nat Biotechnol 32:267–273. https://doi.org/10.1038/nbt.2800

17. Parnas O, Jovanovic M, Eisenhaure TM et al (2015) A genome-wide CRISPR screen in primary immune cells to dissect regulatory networks. Cell 162:675–686. https://doi.org/10.1016/j.cell.2015.06.059

18. Steinhart Z, Pavlovic Z, Chandrashekhar M et al (2016) Genome-wide CRISPR screens reveal a Wnt–FZD5 signaling circuit as a druggable vulnerability of RNF43-mutant pancreatic tumors. Nat Med 23:60–68. https://doi.org/10.1038/nm.4219

19. Wang T, Yu H, Hughes NW et al (2017) Gene essentiality profiling reveals gene networks and synthetic lethal interactions with oncogenic ras. Cell 168:890–903.e15. https://doi.org/10.1016/j.cell.2017.01.013

20. Hart T, Brown KR, Sircoulomb F et al (2014) Measuring error rates in genomic perturbation screens: gold standards for human functional genomics. Mol Syst Biol 10:733–733. https://doi.org/10.15252/msb.20145216

21. Hart T, Moffat J (2016) BAGEL: a computational framework for identifying essential genes from pooled library screens. BMC Bioinformatics 17:164. https://doi.org/10.1186/s12859-016-1015-8

Chapter 16

In Vitro Assays for Screening Small Molecules

Ashley A. Adile, David Bakhshinyan, Chitra Venugopal, and Sheila K. Singh

Abstract

Traditionally anti-cancer therapeutics have been designed to target rapidly proliferating cells causing DNA damage and inducing apoptosis. However, with the development of the cancer stem cell (CSC) hypothesis, it has been postulated that a rare, slow dividing tumor cell population is able to escape therapy and contribute to tumor relapse and metastasis. The advances in characterization of CSCs across multiple cancer subtypes have allowed for development of targeted therapies using small molecule inhibitors. In this chapter, we describe two in vitro assays measuring proliferation and secondary sphere formation, which have become gold-standard assays to evaluate the effects of targeted therapies against CSCs. Together these assays constitute a rapid, inexpensive, and highly reproducible pipeline for testing small molecule inhibitors prior to more resource demanding in vivo studies.

Key words Small molecule inhibitors, Targeted therapies, Proliferation, Self-renewal

1 Introduction

At the core of the cancer stem cell (CSC) hypothesis, lies the notion of a rare population of tumor cells capable of driving tumor initiation, progression, and metastasis. Since their identification in both liquid and solid tumors, CSCs have been shown to have an increased self-renewal potential and the ability to power tumor growth through generation of more differentiated tumor cells, which have markedly limited self-renewal potential. The clinical relevance of CSCs comes from their ability to evade even the best chemoradiotherapy regiments available to drive tumor relapse and metastasis, which are associated with the worse clinical outcome. Current anti-cancer therapies have a tendency to kill highly proliferative tumor cells constituting the majority of tumor cells, leaving the slow dividing, CSC population untreated. It is possible that development of strategies targeting both rapidly dividing cells, as well as slow dividing CSCs will result in reduced rates of recurrence and improve overall patient outcome across many cancer subtypes.

Sheila K. Singh and Chitra Venugopal (eds.), *Brain Tumor Stem Cells: Methods and Protocols*, Methods in Molecular Biology, vol. 1869, https://doi.org/10.1007/978-1-4939-8805-1_16, © Springer Science+Business Media, LLC, part of Springer Nature 2019

Since CSCs are defined solely by their functional properties, it is important for researchers to utilize assays evaluating proliferative and self-renewal capacities to accurately describe effects of novel therapies against CSCs.

In this chapter, we describe two robust in vitro assays that can be used to evaluate changes in proliferation and self-renewal capacity of brain tumor CSCs. The proliferative potential of the cells can be estimated through their rate of aerobic respiration by utilizing an oxidation-reduction indicator, resazurin [1]. Once resazurin enters a viable cell, it is irreversibly reduced to its highly fluorescent form, resorufin. The measured fluorescence intensity can then be accurately correlated with the viability of the cells treated with a therapeutic agent. Unlike chemical-based proliferation assay, secondary sphere formation assays rely on visual quantification of spheres formed from a single cell suspension. The optimization of the assay measuring the changes to self-renewal requires optimization of multiple aspects, including growth conditions, plated cell density, number of days prior to assay read out, and number of cells constituting a sphere.

The two assays described in this book chapter represent a robust and inexpensive way to accurately evaluate in vitro effects of novel therapies against CSCs, prior to more extensive in vivo preclinical validation experiments.

2 Materials

Ensure that all materials are sterile and have been stored in accordance with the manufacturer's guidelines. Prepare reagents at room temperature in a Class II Biological Safety Cabinet (BSC) with the proper Personal Protective Equipment (PPE).

1. Appropriate cell culture medium (*see* **Note 1**). Store at 4 °C. Prior to experimentation, pre-warm media to 37°C in water bath or incubator.

2. Sterile 96-well microwell plates.

3. Tissue culture plates.

4. 1.5 mL Eppendorf tubes.

5. HERAcell 150 incubator that maintains a temperature of 37°C and 5% CO_2 levels.

6. 15 mL and 50 mL centrifuge tubes.

7. 75 mm tube with 35 μm-cell strainer cap.

8. Staining Buffer: 2 mM Phosphate-buffered saline (PBS) with Ethylenediaminetetraacetic acid (EDTA). Add 40 μL of EDTA to 10 mL of 1× PBS (without Ca^{2+} and Mg^{2+}). Store at 4 °C.

9. Viability Dye: 1 mg/mL 7-aminoactinomycin D (7-AAD). Weigh and dissolve 1 mg of 7-AAD powder in 50 mL of absolute methanol prior to adding 950 mL of 1× PBS based on the protocol described in Schmid et al. [2] (*see* **Note 2**). Store at 4 °C, protected from light source.

10. Flow Cytometer.

11. Countess II FL Cell Counter and slides.

12. Trypan Blue Stain (0.4%). Store at room temperature.

13. Dimethyl sulfoxide (DMSO): commonly used as vehicle to dissolve drugs and negative control.

14. 10 mM stock of small molecule inhibitor.

15. PrestoBlue Cell Viability Reagent (*see* **Note 3**). Store at 4 °C.

16. Microplate Analysis Equipment and Software: FLUOstar Omega Microplate Reader with excitation and emission wavelengths of 540 nm and 590 nm, respectively and MARS data analysis software.

---

3   Methods

3.1   Cell Suspension Preparation and Plating

1. Collect and centrifuge cell cultures in accordance with the established standard laboratory culture protocols. Remove the supernatant. Depending on resources available and cell culture, automated (flow cytometry) or manual cell plating can be used.

   (a) *Automated cell plating.*

      1. Resuspend cell pellet in 1 mL of 2 mM PBS-EDTA (*see* **Note 4**).

      2. Filter solution into a 75 mm Tube with a 35 μm cell strainer cap. Add 10 μL of 7-AAD.

      3. In a 96-well microwell plate, add 150 μL of the appropriate cell culture media to the wells designated for proliferation assay or 200 μL to the wells designated for self-renewal assay. Samples for each drug dilution should be plated in quadruplicates.

      4. Using flow cytometer, sort 1000 live cells per well designated for proliferation assay or 200 live cells per well designated for self-renewal assay. For each drug tested, ensure at least one well is left blank; it will be used as a negative control during the analysis.

   (b) *Manual cell plating.*

      1. Resuspend cell pellet in 1 mL of the appropriate cell culture media.

      2. Obtain Countess II FL Cell Counter slide, an Eppendorf tube and Trypan Blue Stain (0.4%). Pipet 10 μL of

Trypan Blue Stain (0.4%) and 10 μL of cell solution into the Eppendorf tube. Thoroughly mix and load 10 μL of the cell suspension-Trypan Blue solution into the Countess II FL Cell Counter slide.

3. Determine the volume of cell suspension containing the required number of cells (*see* **Note 5**).

4. Pipet the calculated volume into each well of a 96-well microwell plate. Samples are plated in quadruplicates for each drug tested. Designate a blank well, with 200 μL of culture medium. Add the appropriate cell culture media to each used well to reach a total volume of 150 μL per well designated for proliferation assay or 200 μL per well designated for self-renewal assay.

3.2 Drug Dilution Preparation

1. Allow cells to settle in HERAcell 150 incubator for 4 h (*see* **Note 6**).

2. For each drug tested, obtain the appropriate cell culture media, DMSO, 10 mM drug aliquot and 12 Eppendorf tubes labeled from 1 to 12.

3. Add 20 μL of DMSO into Tube 1 and 10 μL into Tubes 2 to 10 (Fig. 1).

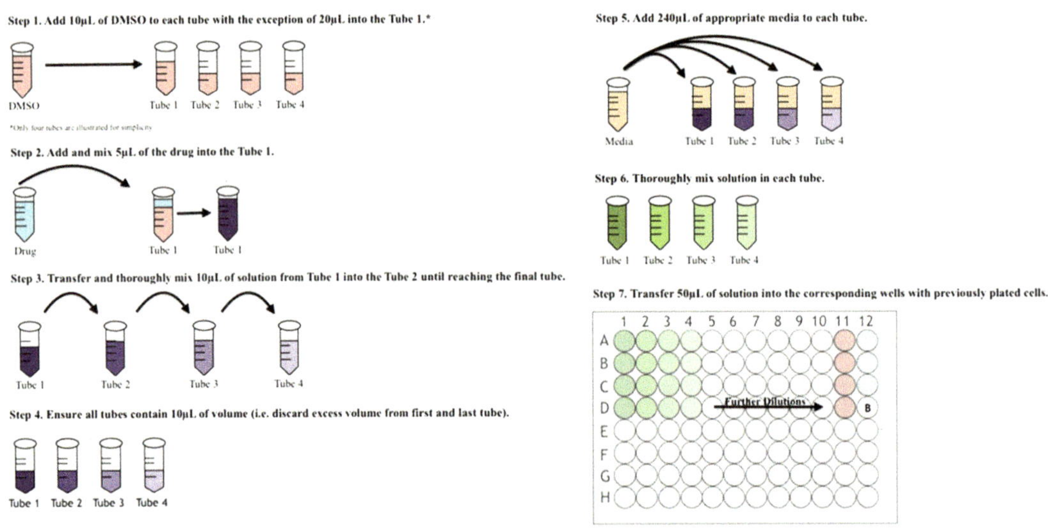

Fig. 1 Schematic depicting twofold serial drug dilutions within in vitro proliferation assay. Cells are plated into half of a 96-well microwell plate with at least one blank well, via flow cytometry or manual cell counting. DMSO is added to each tube, in which Tube 1 contains double this amount. 10 mM of drug is also added to Tube 1. Solution is transferred into the next tube, until reaching the final tube. Once the excess volume in the first and last tubes are discarded, the appropriate cell culture media is added to all tubes. Each drug solution is plated in quadruplicates in their specific columns, while the last two columns of the plate are intended for DMSO and untreated cells, respectively. All steps involve thorough drug resuspension

4. Add and thoroughly mix 5 μL of the 10 mM drug aliquot to Tube 1.

5. Transfer and thoroughly mix 10 μL of the drug solution from Tube 1 into Tube 2.

6. Transfer and thoroughly mix 10 μL of the drug solution from Tube 2 into Tube 3.

7. Repeat **steps 5** and **6** until reaching Tube 10 (Fig. 1).

8. Discard excess volume in Tube 1 and 10 (i.e., 5 μL from Tube 1 and 10 μL from Tube 10).

9. Resuspend the 10 mM DMSO aliquot and transfer 10 μL into Tube 11.

10. Add 10 μL of the appropriate culture cell media into Tube 12.

11. Add and thoroughly mix 240 μL of the appropriate culture cell media into Tubes 1 to 12.

12. Transfer 50 μL of solution from each tube into the respective wells of the prepared 96-well mircowell plate.

3.3 Proliferation Assay

1. Incubate drug treated cells for 3–7 days depending on the proliferative potential of the cell culture.

2. Add 20 μL of PrestoBlue Cell Viability Reagent to each used well of the 96-well microwell plate, including the well designated as "blank."

3. Incubate cells for 3–6 h (*see* **Note 7**).

4. Insert 96-well microwell plate into FLUOstar Omega Microplate Reader to measure fluorescence intensity and thus proliferation levels. Ensure excitation wavelength is set to 540 nm and emission wavelength is at 590 nm.

5. Use MARS data analysis software to analyze data. Input data into GraphPad Prism computer software to determine the effective IC_{50} concentration of the tested drug (Fig. 2).

3.4 Self-renewal Assay

1. Incubate cells in HERAcell 150 incubator for 3 days (*see* **Note 8**).

2. Thoroughly examine each well. Count and record the amount of neurospheres for each well.

3. Incubate cells in HERAcell 150 incubator for an additional 4 days.

4. Thoroughly examine each well. Count and record the amount of neurospheres for each well.

5. Input results into GraphPad Prism computer software to generate a bar graph, comparing the secondary sphere formation ability of the cells treated with the drug, relative to the control (Fig. 3).

6. Discard the plate following analysis.

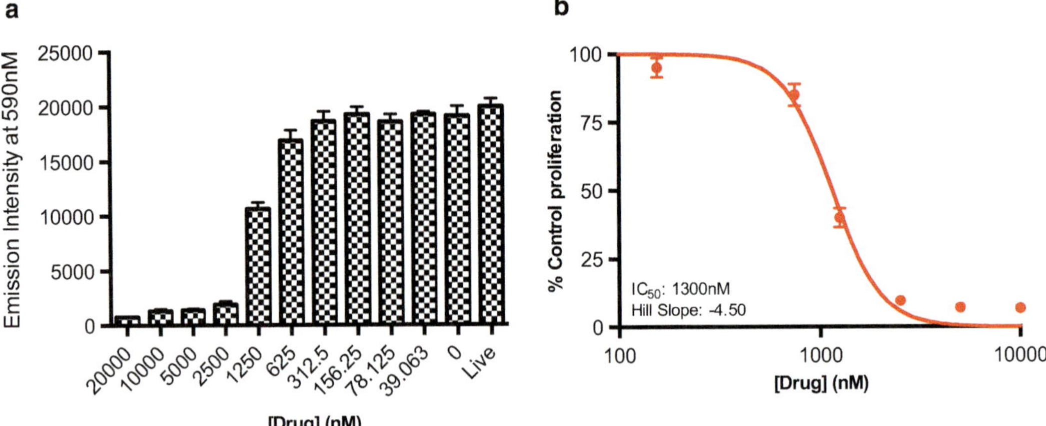

Fig. 2 Determination of effective IC_{50} concentration of a small molecule inhibitor in human cancer cell line. Various versions of the IC_{50} curve are depicted following a 72 h incubation period and addition of PrestoBlue: (**a**) based on emission intensity and (**b**) log2 transformed values of the measured fluorescence intensity. 1000 viable cells/well were sorted into a 96-well microwell plate via flow cytometry, after-which twofold dilutions of a small molecule were prepared and added ($n = 4$)

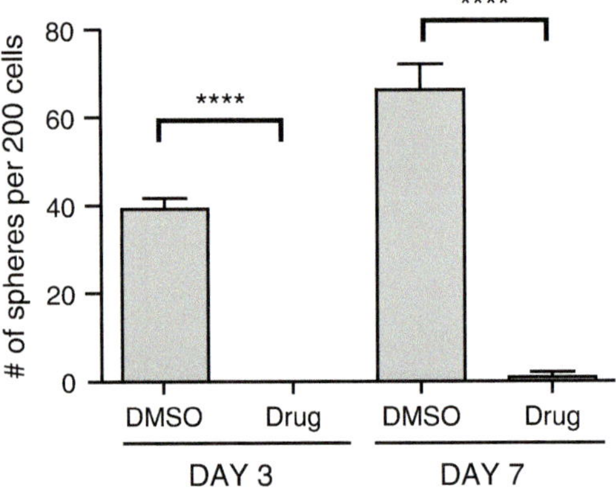

Fig. 3 Certain small molecule inhibitors reduce self-renewal capacity even when dosed in low micromolar range. Cells were treated with IC_{80} concentrations of the drug or DMSO control at a density of 200 cells/well. In vitro sphere formation assay revealed significant decrease in the number of neurospheres formed post treatment. Neurospheres were defined as seven or more tightly clustered cells

4   Notes

1. Human neural stem cell (NSC) lines and tumor samples are cultured in NeuroCultTM NS-A Basal Medium supplemented with 50 mL NeuroCultTM Supplement, 10 µg epidermal growth factor (EGF), 5 µg fibroblast growth factor (FGF), 0.1% heparin and 1% penicillin-streptomycin. Commercially

available cell lines are cultured in accordance with the provider's guidelines.

2. Common viability dyes include LIVE/DEAD® Fixable Near-IR Dead Cell Stains and 7-AAD. Both are used to identify non-apoptotic cells; however, traditional nucleic acid dyes, such as 7-AAD, are able to penetrate compromised membranes and directly bind to DNA [3]. This allows 7-AAD to be used on unfixed, live cells and is therefore preferred over the LIVE/DEAD® Fixable Near-IR Dead Cell Stains that are specifically for fixed cells [4]. Once added, incubate cells for 20 min on ice for LIVE/DEAD® Fixable Near-IR Dead Cell Stains or 15 min at room temperature for 7-AAD. Ensure solutions are kept away from light source.

3. Turn off Class II BSC light while working with PrestoBlue Cell Viability Reagent, due to reagent's light sensitivity.

4. The amount of PBS-EDTA and subsequently the viability dye (i.e., 7-AAD) added is dependent on the size of the cell pellet. For larger pellets, it is recommended to resuspend in 1 mL of PBS-EDTA and add 20 µL of 7-AAD per sample (1×10^6 cells), while preserving 2% 7-AAD in staining buffer for smaller pellets [5]. We find that 1% 7-AAD: PBS-EDTA is sufficient [6, 7].

5. While flow cytometry relies on highly computerized system, manual viable cell counting methods have demonstrated similar accuracy [8]. The following formula can be used for determining volume of cell suspension to be added into each well:

$$\frac{\text{Number of cells required}\left(\frac{\text{cells}}{\text{well}}\right)}{\text{Cell concentration}\left(\frac{\text{cells}}{\text{mL}}\right)} = \text{Volume of cell suspension required}\,(\text{mL/well})$$

6. We find that a 4 h incubation period allows for optimal results of this assay, as cells remain undisturbed, settle and grow. This may vary with cell morphology (adherent versus cells in suspension) and standard laboratory procedures.

7. We find a 4 h incubation period with PrestoBlue Cell Viability Reagent is sufficient.

8. An in vitro secondary sphere formation assay can be performed in conjunction with a limiting dilution assay (LDA). We define spheres as seven or more cells in contact with each other to which we count three and 7 days after initial treatment of either the drug or DMSO control, in replicates of four. Here, the secondary sphere formation assay involves 200 cells/well, but deviations from this may occur with overly confluent cell lines, such that 100 cells/well is also sufficient.

References

1. Anoopkumar-Dukie S, Carey JB, Conere T et al (2005) Resazurin assay of radiation response in cultured cells. Br J Radiol 78:945–947

2. Schmid I, Uittenbogaart C, Jamieson BD (2007) Live-cell assay for detection of apoptosis by dual-laser flow cytometry using Hoechst 33342 and 7-amino-actinomycin D. Nat Protoc 2:187–190

3. Arndt-Jovin DJ, Jovin TM (1989) Fluorescence labeling and microscopy of DNA. Methods Cell Biol 30:417–448

4. Perfetto SP, Chattopadhyay PK, Lamoreaux L et al (2006) Amine reactive dyes: an effective tool to discriminate live and dead cells in polychromatic flow cytometry. J Immunol Methods 313:199–208

5. Schmid I, Krall WJ, Uittenbogaart CH et al (1992) Dead cell discrimination with 7-amino-actinomycin D in combination with dual color immunofluorescence in single laser flow cytometry. Cytometry 13:204–208

6. Singh SK, Clarke ID, Terasaki M et al (2003) Identification of a cancer stem cell in human brain tumors. Cancer Res 63:5821–5828

7. Singh SK, Hawkins C, Clarke ID et al (2004) Identification of human brain tumor initiating cells. Nature 432:396–401

8. Cadena-Herreraa D, Esparza-De Larab JE, Ramírez-Ibañezb ND et al (2015) Validation of three viable-cell counting methods: Manual, semi-automated, and automated. Biotechnol Rep 7:9–15

Chapter 17

Drug Delivery in an Orthotopic Tumor Stem Cell-Based Model of Human Glioblastoma

Elena Binda, Alberto Visioli, Nadia Trivieri, and Angelo Luigi Vescovi

Abstract

Grade IV gliomas, also known as glioblastoma multiforme (GBM), are incurable, lethal brain tumors, whose average life expectancy is around 15 months. There is a desperate need for a better understanding of the basic biology of these tumors, in order to devise novel, more specific and effective therapeutics. The handling of GBM represents a daunting challenge to clinicians, also considering the few therapeutic options available, none of which can significantly alter the inevitable lethal outcome of these tumors. Hence, the development of effective therapies would greatly benefit from the availability of in vivo GBM models that can reliably mimic the characteristics of malignant cells and the features of the human disease. Candidate new drugs have to be tested in these in vivo models by adopting settings concerning direct intra-brain delivery in order to define their overall therapeutic efficacy under clinical-like conditions. Here, we describe local intracranial delivery of drugs by osmotic mini-pumps.

Key words Glioblastoma multiforme, Glioblastoma stem cells, Animal models, Drug delivery, Osmotic mini-pumps

1 Introduction

The protective environment of the central nervous system (CNS) poses significant obstacles to the treatment of malignant brain tumors since, very often, systemic administration of anti-neoplastic agents by oral or intravenous routes fails to achieve effective drug concentrations at the tumor site, even at toxic systemic doses [1]. The presence of biological barriers including the blood-brain barrier (BBB), blood-cerebrospinal fluid barrier (BCSFB), and the blood-tumor barrier (BTB) [2], as well as the presence of drug-metabolizing enzymes in cerebral micro-vessels [3] further prevents accumulation of adequate drug levels at the appropriate site [1, 2, 4]. We were able to overcome these obstacles by means of intracranial sustained delivery using osmotic mini-pumps (Alzet micro-osmotic pump, Durect Corp.) (Fig. 1) [5, 6], a method

Sheila K. Singh and Chitra Venugopal (eds.), *Brain Tumor Stem Cells: Methods and Protocols*, Methods in Molecular Biology, vol. 1869, https://doi.org/10.1007/978-1-4939-8805-1_17, © Springer Science+Business Media, LLC, part of Springer Nature 2019

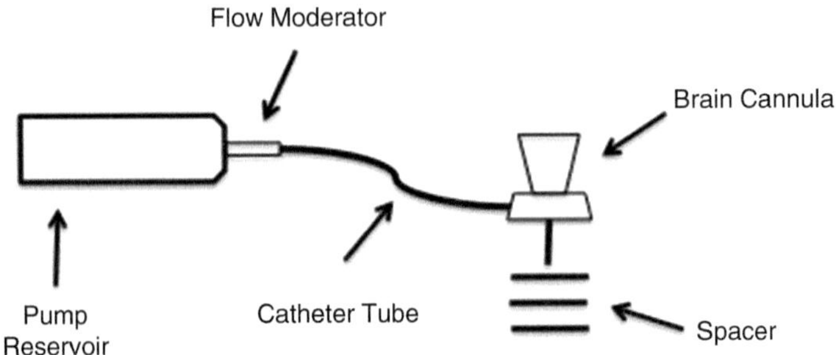

Fig. 1 Schematic of osmotic mini-pump assembling as described in Subheading 3.7

analogous to the intracranial, convention enhanced delivery (CED) used in GBM patients [7–9].

Glioblastoma stem cells (GSCs) were injected into the nucleus striatum of SCID mice by stereotaxic implantation and two experimental paradigms were used: (1) sustained delivery in vivo by osmotic mini-pumps occurred immediately thereafter (co-treatment) or (2) allowed to establish sizeable tumors, i.e., 11 days, before beginning treatment through continuous intracranial infusion (post-treatment). In order to gather both quantitative and broader, more homogenous data, in vivo experiments were carried out by transplanting firefly luciferase-tagged (F-Luc) GSCs, using an Ivis Lumina Xenogen for imaging analysis (Fig. 2) [5, 6]. By these approaches, the compounds of choice were delivered for up to two continuous weeks in mice, traveling through the brain parenchyma based primarily on diffusion [10]. We have developed significant expertise to be able to warrant effective delivery of drugs at the specific site of choice.

2   Materials

2.1   Micro-surgery Tools

1. Stereotaxic frame (*see* **Note 1**).
2. Cannula holder.
3. Spring scissors—10 mm—blades angled.
4. Spring scissors—8 mm—blades straight.
5. Iris scissors—delicate straight 11.5 cm.
6. Iris scissors—delicate angled side.
7. Surgical scissors—straight 14.5 cm.
8. Providence hospital hemostat—straight.
9. Offset bone nipper.
10. Semken forceps—straight 13 cm.
11. Semken forceps—curved 13 cm.

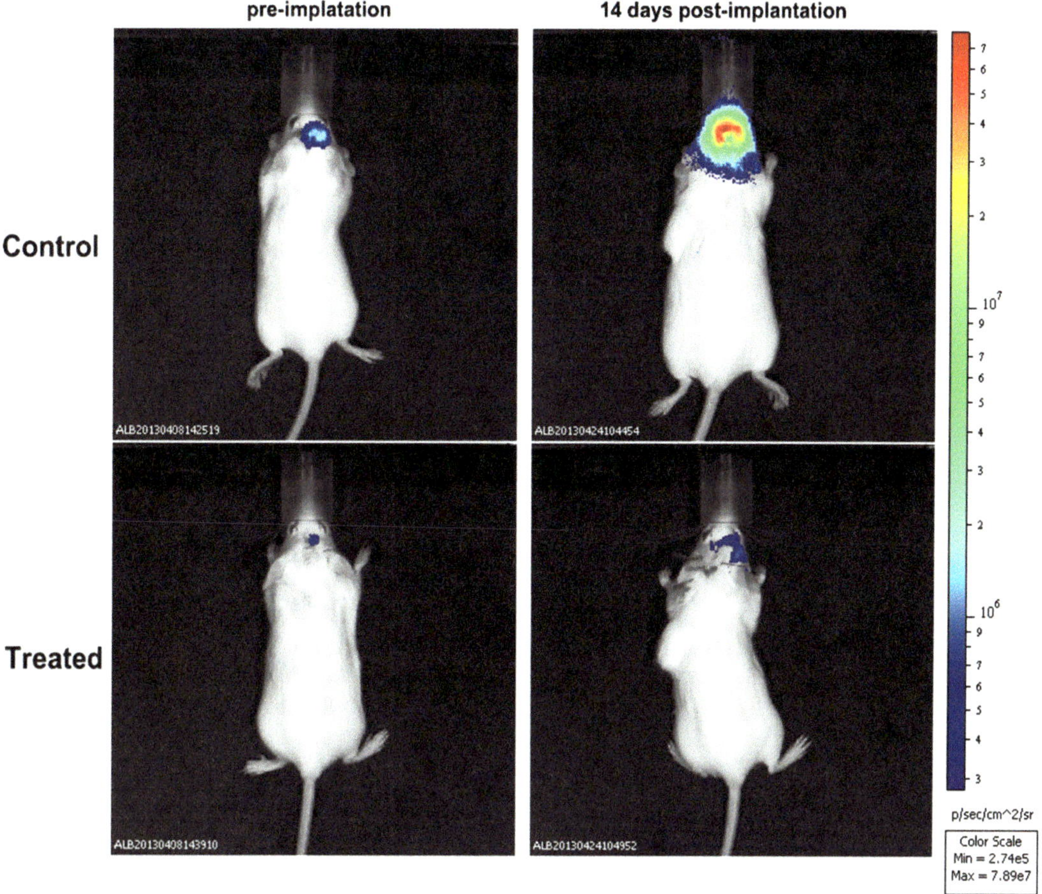

Fig. 2 Imaging of luciferase-tagged GSCs injected into the brain of *SCID* mice in a post-treatment experiment. The efficacy of drug delivery is shown by the inhibition of tumor growth in treated mice as compared to controls

12. Bulldog serrefine—straight 50 mm.

13. Dumont #5–45 forceps—dumoxel.

14. Dumont #55 mirror finish forceps—inox.

15. Dumont #7 forceps—dumoxel.

16. Tissue scissors blunt/blunt.

17. Friedman-pearson roungeur—1 mm cup.

18. Tissue forceps—1 × 2 teeth 12 cm.

19. Graefe forceps—0.8 mm tips curved.

20. Scalpel number 11.

21. Micro-drill with 0.5 mm steel burrs.

22. Sterile phosphate-buffered saline (PBS) or sterile 0.9% saline solution.

23. Antiseptic disinfectant.

24. Absorbable sutures 4–0 size.

25. Non-absorbable sutures 3–0 size.

26. Syringe with needle 25 G × 5/8″ (for intra-peritoneal anesthesia).

27. Micro-syringe (5 µL, 30 G needle).

2.2 Animal Anesthesia

(*See* **Note 2**).

For injectable anesthesia:

Prepare 10 mL of an anesthetic cocktail solution (80 mg/kg body weight Ketamine and 10 mg/kg body weight Xylazine):

1. Mix 0.8 mL of Ketamine (100 mg/mL), 0.5 mL of Xylazine (20 mg/mL), and 8.7 mL of sterile 0.9% saline solution.

2. Inject 0.1 mL every 10 g of body weight intra-peritoneally.

For inhalant anesthesia:

1. Isofluorane 3–5% for the induction of anesthesia.

2. Isofluorane 1–3% for the maintenance of anesthesia.

2.3 GSCs Subculture Material

1. Plasticware (flasks, polypropylene conical tube, bottle top filters, syringe filters, pipettes, universal fit filter tips).

2. NeuroCult® NSC Basal Medium.

3. NeuroCult® NSC Proliferation Supplements.

4. FGF2 (10 ng/mL).

5. EGF (20 ng/mL).

6. Dulbecco's Modified Eagle Medium 1× (DMEM).

2.4 Animal Analgesia

1. After every surgical procedure, Flunixin meglumine (2 mg/kg) has to be injected subcutaneously every 12–24 h for 5 days.

2.5 Osmotic Mini-Pumps

1. Brain Infusion Kit 3 (Alzet micro-osmotic pump, Durect Corp).

2. Pump reservoir (*see* **Note 3**).

3. Syringe (0.3 mL) with 30 G × 8 mm needle.

4. Sterile 0.9% saline solution.

5. Cyanoacrylate adhesive.

3 Methods

3.1 GSCs Subculturing and Procedure for Orthotopic Injection

1. Tap sides of flasks to dislodge spheres to be injected and transfer content of the flask to 15 mL sterile conical tubes using a sterile plastic pipette. Use 5 mL fresh growth medium to rinse flask and add rinse to the tube. Subculture when the

neurospheres reach roughly 150–200 mm diameter, this will require approximately 5–7 days for GSC cells.

2. Pellet cell suspension by centrifugation at 192 × *g* for 10 min. Remove the supernatant leaving behind approximately 250 μL.

3. Use a Gilson 200 set at a volume of 190 μL. Rinse tip with medium first, to avoid cell sticking inside the tip. Slightly press tip against the bottom of the tube gently triturate pellet. Optimal dissociation should be obtained with 40–60 passages through the tip for GSCs cultures. Rinse down sides of tube periodically to dislodge undissociated spheres (*see* **Note 4**).

4. Count viable cells by trypan blue exclusion and seed cells at the appropriate density in NeuroCult NS-A medium containing FGF2 and EGF in untreated tissue flasks: 1.5×10^4 cells/ cm^2 (*see* **Note 5**).

5. For orthotopic injection, repeat steps from 1 to 3.

6. Count viable cells by trypan blue exclusion and add cells at the appropriate density to 1 mL of DMEM in a microcentrifuge tube.

7. Pellet cell suspension by centrifugation at 344 × *g* for 10 min.

8. Resuspend the pellet in DMEM at the appropriate density: for orthotopic xenografts 3 μL of a 1×10^5 cells/μL has to be injected into the right striatum of SCID mice.

3.2 Anesthetic Procedure

1. Place the mouse on a rough surface while holding the tail firmly.

2. Grasp the cape gently and firmly with your free hand and lift the mouse.

3. Insert the needle of the syringe medially into the area between the inside of the thigh and the nipple and then inject the anesthetic mixture solution.

4. Put the mouse in its cage awaiting loss of consciousness due to the absorption of the anesthetic mix.

3.3 Animal Mounting on the Stereotaxic Frame

1. Swing the nose clamp and incisor bar down and out of the way, and then lock one of the ear bars at 3.0 mm from center.

2. While supporting the animal's head, guide the tip of the lock ear bar into the external auditory meatus and hold firmly. Slide the other ear bar into the opposite ear and apply steadily increasing pressure until small popping sounds are heard.

3. Swing the incisor bar up and slide it just into the mouth while holding the jaw open with a pair of blunt forceps. Apply downward pressure to the bridge of the nose while sliding the incisor bar back into the mouth.

4. Apply sterile 0.9% saline solution onto mouse eyes to prevent them for drying out.

<table>
<tr><td valign="top">

3.4 Animal Stereotaxic Surgery

</td><td>

1. Shave and wash the scalp with antiseptic solution. Starting slightly behind the eyes, make a midline sagittal incision about 1.5 cm long and expose the skull.

2. Scrape away the periosteal connective membrane, which adheres to the skull, first with a scalpel and then with a cotton swab.

3. Fit the micro-syringe loaded with cell suspension on the mono-injector of the stereotaxic frame and set the dorsal-ventral (DV), anterior-posterior (AP) and medial-lateral (ML) coordinates to bregma zero (*see* **Note 6**).

4. Fit the micro-syringe to the DV, AP, and ML coordinates and drill a hole through the skull.

5. Insert the needle of the micro-syringe through the hole reaching the DV coordinates and inject cells.

6. After 5 min remove the micro-syringe.

7. Suture the incision and remove the animal from the stereotaxic frame. Recover and place the animal into the cage.

</td></tr>
<tr><td valign="top">

3.5 Osmotic Mini-Pumps Priming Procedure

</td><td>

1. Place the pump in a sterile 0.9% saline solution at 37 °C for at least 6 h prior to implantation.

2. Remove the pump from the saline solution and fill the pump immediately.

</td></tr>
<tr><td valign="top">

3.6 Osmotic Mini-Pumps Filling Procedure

</td><td>

1. Check the instructions for the mean fill volume for the lot of pumps that will be used.

2. Draw up the drug solution with a syringe (0.3 mL) with 30 G × 8 mm needle (*see* **Note 7**).

3. Remove the flow moderator and hold the pump in an upright position.

4. Insert the needle of the syringe through the opening at the top of the pump reservoir until it can go no further.

5. Slowly push the plunger of the syringe and hold the pump in an upright position (*see* **Note 8**).

6. When the solution appears at the outlet, stop filling and carefully remove the needle.

</td></tr>
<tr><td valign="top">

3.7 Preparation of the Brain Infusion Assembly

</td><td>

1. Determine the stereotaxic coordinates for the target infusion site and calculate the desired cannula length (*see* **Note 9**).

2. Cut the catheter tubing to the length determined (1.5 cm approximately). Attach one end of the tubing to the cannula, spreading a thin film of cyanoacrylate for 2 mm on brain infusion cannula, and the other end to the pump flow moderator without the white flange cap.

</td></tr>
</table>

3. Check the attachment by gently pulling on the catheter (*see* **Note 10**).

4. Wipe off the excess solution and insert the flow moderator previously joined to the pump reservoir (*see* **Note 11**).

5. Fill the brain infusion assembly with the solution of the drug to be delivered. To do this, attach a syringe containing this solution to the remaining catheter tubing and connect the tubing to the free end of the flow moderator.

3.8 Application of Osmotic Mini-Pumps

1. Mice have to be anesthetized with intraperitoneal injection of an appropriate dose of ketamine/xylazine mixture. .

2. Shave and wash the scalp with an antiseptic solution.

3. Starting slightly behind the eyes, make a midline sagittal incision about 1.5 cm long and expose the skull. With a scalpel, lightly scrape the exposed skull area and pat it dry (*see* **Note 12**).

4. Identify the bregma and with this reference determine the coordinates for the intrathecal drug infusion.

5. Drill a hole through the skull.

6. Remove from stereotaxic frame the monoinjector and insert the cannula holder.

7. Insert the L-shaped cannula, previously attached to the delivery system, through the skull. Prepare a subcutaneous pocket in the midscapular area of the back of the mouse to receive the osmotic mini-pump (*see* **Note 13**).

8. Insert the tab on the top of the cannula to the cannula holder and then through the hole in the skull.

9. Spread cyanoacrylate adhesive under the cannula and then placed it to the skull.

10. After at least 5 min close the scalp with interrupted sutures.

11. Remove the animal from the stereotaxic frame. Recover and place the animal into the cage.

4 Notes

1. Stereotaxic is a versatile device used to localize precisely an area in the brain by means of coordinates related to intracerebral structures. Heterogeneity among different kinds of devices is related to the resolution of the Vernier scale (10–100 μm) and to the manual or digital configuration of the manipulators.

2. For the in vivo procedures with mice both injectable and inhalant anesthesia can be used. Ketamine, which can be included in the first one, may be combined with the alpha-

2 agonist named xylazine in the same syringe to produce a deep level of sedation. In some situations and in some species and strains, an adequate depth of anesthesia for surgery may be attained. Advantages of ketamine/alpha2-agonist combinations are that they may be combined in one syringe, that they may produce short-term surgical anesthesia with good analgesia, and that recovery can be hastened by reversing the alpha2-agonist. If a ketamine/alpha2-agonist combination is used for surgery longer than 20 min, animals will likely require additional anesthetic. Re-dosing with a lower dose of ketamine rather than the combination is usually safer, as the cardiovascular depression of alpha2-agonists is often longer-lasting than the sedation or analgesia produced. However, re-dosing repeated with ketamine alone will not produce a surgical plane of anesthesia. Preferably, a gas anesthetic should be used to continue the anesthetic for procedures longer than 30 min; this is a safer alternative. In this case it is necessary to consider the use of the accessory for maintaining anesthesia when the animal is placed on the stereotaxic frame [11].

3. In choosing the appropriate model of osmotic mini-pump consider the different release rates and durations.

4. Do not dissociate neurospheres when they are too small (the yield will be low), nor leave neurospheres overgrow (dark region in the core of the sphere). In this case the number of dead cells inside the spheres will be high, trituration will be difficult, and viability of the culture will be very low.

5. The total cell number should increase twofold at each passage for GSCs. Viability after dissociation should never fall below 50–60% of the total cells.

6. Stereotaxic surgery is a well-established technique for localizing specific structures within the brain of living animals. The bony features on the surface of the skull have to be used as reference point for determining the location of brain structures. These features are named lambda and bregma. The latter one is the intersections of the sagittal suture with the coronal suture. Use mouse stereotaxic brain atlas (Paxinos and Francklin) to determine the AP, ML, and DV coordinates of brain structures.

7. The syringe has to be free of air bubbles. Allow extra syringe volume for spillage.

8. Rapid filling of mini-osmotic pumps should be avoided because it can introduce air bubbles into the reservoir.

9. As described by the manufacturer's instructions, the L-shaped cannula included in the Brain Infusion Kit 3 will penetrate approximately 3 mm below the surface of the skull. To obtain

shorter cannula, spacers, included in the Brain Infusion Kit, can be used.

10. As recommended by the manufacturer's instructions, the distance between the location at which the cannula will be placed and the site of pump implantation has to be measured. The catheter that connects the cannula to the pump reservoir should be 25% longer than the distance between the subcutaneous site of the pump and the location of the cannula, to allow free movement of the animal's head and neck.

11. The insertion of the flow moderator will displace some of the solution from the filled pump out of brain infusion needle. The flow moderator must be fully inserted into the body of the pump.

12. Scraping should remove the periosteal connective tissue, which adheres to the skull, permitting a good adhesion of the cyanoacrylate adhesive.

13. The pocket has to be created by opening and closing a surgical forceps to blunt dissect a short subcutaneous tunnel from the scalp incision to the mid-scapular area. The pocket should be large enough to accommodate the pump and permit some pump movement, but not so large as to allow the pump to slip down onto the flank of the animal.

References

1. Rautioa J, Chikhale PJ (2004) Drug delivery systems for brain tumor therapy. Curr Pharm Des 10:1341–1353

2. Lesniak MS, Brem H (2004) Targeted therapy for brain tumours. Nat Rev Drug Discov 3:499–508

3. Ghersi-Egea JF, Leininger-Muller B, Cecchelli R, Fenstermacher JD (1995) Blood-brain interfaces: relevance to cerebral drug metabolism. Toxicol Lett 82-83:645–653

4. Kusuhara H, Sugiyama Y (2001) Efflux transport systems for drugs at the blood-brain barrier and blood-cerebrospinal fluid barrier (Part 2). Drug Discov Today 6:206–212

5. Binda E, Visioli A, Giani F, Lamorte G, Copetti M, Pitter KL et al (2012) The EphA2 receptor drives self-renewal and tumorigenicity in stem-like tumor-propagating cells from human glioblastomas. Cancer Cell 22:765–780

6. Binda E, Visioli A, Giani F, Trivieri N, Palumbo O, Restelli S et al (2017) Wnt5a drives an invasive phenotype in human glioblastoma stem-like cells. Cancer Res 77:996–1007

7. Bobo RH, Laske DW, Akbasak A, Morrison PF, Dedrick RL, Oldfield EH (1994) Convection-enhanced delivery of macromolecules in the brain. Proc Natl Acad Sci U S A 91:2076–2080

8. Burgess A, Hynynen K (2014) Drug delivery across the blood-brain barrier using focused ultrasound. Expert Opin Drug Deliv 11:711–721

9. Debinski W, Tatter SB (2009) Convection-enhanced delivery for the treatment of brain tumors. Expert Rev Neurother 9:1519–1527

10. Theeuwes F, Yum SI (1976) Principles of the design and operation of generic osmotic pumps for the delivery of semisolid or liquid drug formulations. Ann Biomed Eng 4:343–353

11. Fish RE, Brown MJ, Danneman PJ, Karas AZ (2008) American college of laboratory animal medicine. Anesthesia and analgesia in laboratory animals, 2nd edn. Academic Press, San Diego

Chapter 18

Engineering Inducible Knock-In Mice to Model Oncogenic Brain Tumor Mutations from Endogenous Loci

Jon D. Larson and Suzanne J. Baker

Abstract

To maximize the physiological relevance of in vivo brain tumor mouse models designed to study the downstream effects of oncogenic mutations, it is important to express the mutated genes at appropriate levels, in relevant cell types, and in the proper developmental context. For recurrent mutations found in the heterozygous state in tumors, expression of the mutation from the endogenous locus is a more physiologically relevant recapitulation of the brain tumor genome. Here, we describe an approach to generate knock-in mice with an inducible mutation recombined into the endogenous locus. In these engineered mice, the mutated allele is designed for expression controlled by the endogenous promoter and regulatory elements after Cre recombinase-mediated deletion of a *loxP*-STOP-*loxP* cassette inserted upstream of the translational start site. To preserve the structure of the endogenous locus, mutations or additional elements may need to be inserted at a considerable distance from the *loxP*-STOP-*loxP* cassette. We used recombineering to build a construct with two selectable markers and multiple genetic alterations that can be introduced into the endogenous allele in *cis* with a single ES cell targeting.

Key words Recombineering, Knock-in, Transgenic, Mutation, Oncogene, Mouse, Cre recombinase, *loxP*, Brain tumor, Neuro-oncology

1 Introduction

Genetically engineered mouse models are extremely useful tools for the study of normal brain development and neuropathologies including brain tumors. Diverse approaches are available to generate transgenic mice expressing mutated oncogenes in the brain, with varying levels of regulation. Mutated genes may be introduced into a small number of somatic cells through injection of virus, or in vivo delivery of DNA constructs [1, 2]. While expeditious, the oncogenes are not expressed under the physiological regulation conferred by the endogenous promoter and may not reflect a heterozygous state if desired. The small number of cells targeted with such approaches may limit brain tumor development, showing greatest success with very strong oncogenic drivers that may not be

Sheila K. Singh and Chitra Venugopal (eds.), *Brain Tumor Stem Cells: Methods and Protocols*, Methods in Molecular Biology, vol. 1869, https://doi.org/10.1007/978-1-4939-8805-1_18, © Springer Science+Business Media, LLC, part of Springer Nature 2019

most representative of mutations in human tumors. Furthermore, injection-based delivery poses varying technical challenges in different anatomic locations and developmental stages. Germline mutations may highlight connections between normal development and functions in tumorigenic transformation. Many cancer-associated mutations cause embryonic lethality, precluding the use of germline mutations to study brain tumor development. Multiple approaches are used to successfully introduce mutations into the mouse germline such as random integration of transgenic constructs by pronuclear injection (i.e., transgenes) and genome editing using tools such as CRISPR/Cas9-mediated genome editing in zygotes or embryonic stem (ES) cells [3–6]. Genome editing is currently most successful for simple gene knockouts or constitutive mutations, but remains more challenging for complex designs in which multiple elements must be reliably introduced in *cis* within a single allele, particularly through the introduction of large segments of DNA [7, 8]. Currently, more precise control over induction of a particular mutation in specific cell types and at diverse developmental stages can be achieved by Cre recombinase-inducible germline alleles.

Recombineering uses homologous recombination through gap repair in bacteria to genetically introduce alterations in plasmids, and is a powerful method to construct complex vectors to create knock-in alleles with multiple genetic features [9]. As an example, we discuss the design to generate a Cre recombinase-inducible K27M mutation knocked-in to the endogenous *H3f3a* gene to model mutations found in human diffuse intrinsic pontine glioma. We detail the protocol to recover the genomic region of interest from a bacterial artificial chromosome (BAC), introduce a *loxP*-STOP-*loxP* cassette to render the gene Cre recombinase-inducible, and introduce additional mutations or genomic elements and selectable markers. The final knock-in targeting construct is electroporated into mouse ES cells. Designing the knock-in construct with two selectable markers allows simultaneous engineering of multiple alterations in a single ES targeting experiment. The targeted ES cells described in this protocol would be used to generate mice [10] which can be bred with mice expressing Cre recombinase in a tissue-specific and/or inducible manner. The resulting models provide spatial and temporal regulation of the engineered knock-in oncogenic mutation under control of the endogenous promoter and regulatory elements.

2   Materials

2.1   Recombineering

1. BAC containing the genomic locus of the gene of interest, ideally from the same mouse strain as the ES cells that you will use (*see* **Note 1**). For this example, mouse *H3f3a* BAC in

chloramphenicol-resistant plasmid pBACe3.6 (Source BioScience, #bMQ164i21).

2. Recombineering bacterial strain SW102 (NCI, https:// ncifrederick.cancer.gov/research/brb/reagents/ recombineeringReagent.aspx). A modified DH10B strain derived from DY380 [9] with the *galactokinase* (*galK*) gene deleted from the galactose operon for *galK* selection. These cells express bacteriophage λ *Red* genes *exo*, *bet* and *gam* from a temperature-responsive promoter that facilitate gap repair homologous recombination. The gap repair genes are repressed at 30 °C and expressed at 42 °C. These cells are tetracycline resistant.

3. Recombineering bacterial strain SW105 (NCI, https:// ncifrederick.cancer.gov/research/brb/reagents/ recombineeringReagent.aspx). Similar to SW102, derived from EL250 [9] but contains an arabinose-inducible *araC*-P_{BAD} promoter expressing enhanced Flippase (Flpe) and carries no antibiotic resistance.

4. Recombineering plasmid PL451 (NCI, https://ncifrederick. cancer.gov/research/brb/reagents/recombineeringReagent. aspx). Contains *Neomycin* (*Neo*) CDS expressed by eukaryotic phosphoglycerate kinase (*PGK*) and prokaryotic *Em7* promoters [9] to confer G418 resistance in mammalian cells and kanamycin resistance in bacteria, respectively. This cassette is flanked by two Flp recombinase target (*Frt*) sites and one *loxP* site.

5. Recombineering plasmid PL451-no *loxP* derived from PL451. *loxP* site present in original PL451 has been removed to be compatible for use with *loxP*-STOP-*loxP* cassette.

6. Recombineering plasmid p*FrtF3-Neo-FrtF3*. Derived from PL451-no *loxP* (2.1–5 above). This cassette is flanked by two Flp recombinase target (*Frt*) sites containing F3 spacer mutation [11]. This design allows Flp to direct deletion between matched *FrtF3* sites, but helps prevent cross-recombination between *Frt* and *FrtF3* sites.

7. Ultrapure DNAse/RNAse-free water (Invitrogen).

8. 10% L-(+)-Arabinose (SIGMA) stock solution: 100 mg/mL dissolved in Ultrapure DNAse/RNAse-free water and 0.22 μm sterile-filtered.

9. LoTE (3 mM Tris, 0.2 mM EDTA, 0.22 μm filter sterilized).

10. Dimethyl sulfoxide (DMSO).

2.2 Vector Cloning

1. pBluescript KS (+) plasmid for subcloning.

2. pBR322-DT targeting plasmid equipped to express negative selection diphtheria toxin (DT) in mammalian cells to eliminate

cells that undergo incorrect homologous recombination. Backbone provides ampicillin resistance.

3. pLox-Stop-Lox-TOPO plasmid (*loxP*-STOP-*loxP*) [12]. Contains a cassette flanked by *loxP* sites equipped with an adenovirus splice acceptor-tetrameric SV40 polyadenylation array designed to halt transcription. Adjacent to the splice acceptor in the opposite orientation is a *PGK* promoter-*puromycin* (*Puro*) gene-*PGK* polyadenylation cassette designed to express puromycin in mammalian cells.

4. Luria-Bertani (LB) media: 10 g/L Tryptone, 5 g/L Yeast extract-B, 10 g/L NaCl (MP Biomedicals, #3002-031). Autoclave 25 capsules/1 L purified water.

5. LB (Lennox L) agar: 9.14 g/L Enzymatic digest of casein, 4.57 g/L Yeast extract (low sodium), 4.57 g/L NaCl, 13.72 g/L Bacteriological agar (SIGMA, #L7025). Autoclave 1 tablet/50 mL purified water.

6. Tetracycline hydrochloride (SIGMA) stock solution: 15 mg/mL dissolved in methanol.

7. Chloramphenicol (SIGMA) stock solution: 30 mg/mL dissolved in ethanol.

8. Ampicillin sodium salt (SIGMA) stock solution: 100 mg/mL dissolved in Ultrapure DNAse/RNAse-free water.

9. Kanamycin (SIGMA) stock solution: 50 mg/mL dissolved in Ultrapure DNAse/RNAse-free water.

10. QIAquick Gel Extraction Kit (QIAGEN).

11. QIAEX II Gel Extraction Kit (QIAGEN).

12. QIAGEN Large-Construct Kit (QIAGEN).

13. QIAGEN EB buffer from QIAprep Spin Miniprep Kit (QIAGEN).

14. Various restriction enzymes (New England Biolabs, use high fidelity versions when possible).

15. DNA Polymerase I, Large (Klenow) Fragment (New England Biolabs).

16. Expand High Fidelity PCR System, dNTP Pack (Roche).

17. pCR2.1-TOPO Cloning Kit w/ TOP10 chemically competent cells (Invitrogen).

18. pCR-BluntII-TOPO Cloning Kit w/ TOP10 chemically competent cells (Invitrogen).

19. QuickChange II SDM Kit (Agilent Technologies).

20. MegaX DH10B T1^R Electrocompetent *E.coli* cells (Invitrogen).

21. One Shot Stbl3 Chemically Competent *E.coli* cells (Invitrogen).

22. UltraPure Phenol:Chloroform:Isoamyl Alcohol (P:C:I, 25:24:1) (Invitrogen).

23. Chloroform.

24. Isopropanol.

25. Ethanol.

26. Methanol.

27. 3 M sodium acetate.

2.3 Electroporation

1. Gene Pulser Electroporation Cuvette with 0.1 cm gap (BIORAD).

2. Gene Pulser Electroporation Cuvette with 0.4 cm gap (BIORAD).

3. Gene Pulser Xcell Total System (BIORAD).

4. Gene Pulser II System with Capacitance Extender Plus (BIORAD).

5. S.O.C. Medium: 2% Tryptone, 5% Yeast Extract, 10 mM NaCl, 2.5 mM KCl, 10 mM $MgCl_2$, 10 mM $MgSO_4$, 20 mM glucose (Invitrogen).

2.4 ES Cell Targeting

1. Mouse 129 strain-derived ES.

2. Multi-drug resistant mouse embryonic fibroblasts (MEF DR4) (ATCC, #SCRC-1045).

3. G418 sulfate liquid (Geneticin, 50 mg/mL) (Gibco).

4. Puromycin dihydrochloride (SIGMA) stock solution: 2 mg/mL dissolved in Ultrapure DNAse/RNAse-free water.

3 Methods

3.1 Designing the Inducible Knock-In Targeting Construct

Accurately planning all steps to build the knock-in targeting vector before construction begins is critical. For highest efficiency, distinct antibiotic selection and introduction of new restriction enzyme sites need to be considered and designed for each step.

1. Generate a sequence map of the entire wild-type gene locus. Identify key features including endogenous translational start and stop sites, and the location to introduce the selective mutation (Fig. 1a). Plan the alterations that need to be introduced. Examples from the engineered *H3f3a* locus include a *loxP*-STOP-*loxP* cassette, point mutation, *Frt-Neo-Frt* cassette, and an epitope tag (Fig. 1b). Recombineering allows introduction of elements at any location. However, when possible,

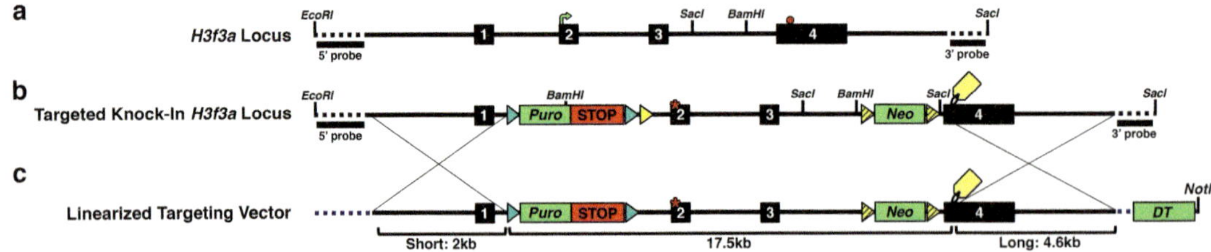

Fig. 1 Designing An Inducible *H3f3a* Knock-In Vector. (**a**) Schematic map of the *H3f3a* gene locus defining intron and exon configuration, endogenous translation initiation (green arrow, exon 2) and stop (red hexagon, exon 4) sites. (**b**) Final design of targeted knock-in *H3f3a* locus illustrating important engineering features including intron 1 placement of *loxP*-STOP-*loxP* (STOP) transcription termination sequence that will also provide puromycin (*Puro*) positive selection in ES cells, exon 2 point mutation (red asterisk), exon 4 in-frame epitope tag (yellow tag) associated with intron 3 placement of *Frt-Neo-Frt* cassette to provide G418 positive selection in ES cells. *loxP* (cyan) *Frt* (yellow) and *FrtF3* (yellow striped) sequences are marked with triangular arrowheads. Dotted line shows genomic locus not included in targeting construct. *EcoRI/BamHI* sites and *SacI* sites generate restriction fragments detected by 5′ and 3′ probes, respectively, to create distinguishable Southern blot DNA fragment sizes before and after targeting, with a *SacI* site designed into the *Frt-Neo-Frt* cassette cloning. Probes are designed to hybridize outside homologous recombination events for proper Southern blot verification. (**c**) Final design of *H3f3a NotI*-linearized targeting vector. Dotted blue line indicates targeting vector backbone. Vector backbone provides Diphtheria toxin (*DT*) negative selection in ES cells. Short (2 kb) and long (4.6 kb) homology arms flank engineered locus (17.5 kb) to guide homologous recombination outside of engineered elements. Not drawn to scale so that elements can be clearly marked

avoid repetitive genomic sequences for recombineering-mediated homologous recombination (*see* **Note 2**). Furthermore, ideally position targeting cassettes (e.g., *loxP*-STOP-*loxP*) at least 200 bp away from splice donors and splice acceptors. Map restriction enzyme sites to assist designing recombineering vectors that will be used for locus retrieval and incorporating engineered elements (Fig. 1a, b).

2. Plan the location for the *loxP*-STOP-*loxP* cassette. For an inducible knock-in, a transcription termination sequence (e.g., *loxP*-STOP-*loxP*) should ideally be placed in an intron or 5′ untranslated region upstream of the endogenous translation initiation codon. To avoid disruption of regulatory elements for your gene of interest or surrounding genes, check the ENCODE database (https://www.encodeproject.org/) to mark potential promoters and enhancers. For *H3f3a* where the canonical translation initiation start is located in exon 2, we designed the *loxP*-STOP-*loxP* cassette to be positioned in intron 1 (Fig. 1b).

3. Plan additional engineered features (e.g., point mutations, epitope tags, etc.). The likelihood of homologous recombination occurring in between the *loxP*-STOP-*loxP* cassette and an additional introduced mutation is directly proportional to genomic distance. For large distances, incorporation of a second selectable marker will greatly enhance the efficiency of

identifying the correctly targeted locus in ES cells. In the *H3f3a* locus, we introduced a K27M point mutation in exon 2, approximately 900 bp from the *loxP*-STOP-*loxP* cassette, which provides puromycin positive selection. If the alterations to be introduced are close together as in this case, the targeting vector could be complete with this single selection. However, we also inserted an in-frame epitope tag to be positioned just upstream of the canonical stop codon in exon 4, approximately 8.7 kb from the K27M point mutation. To allow double selection to maximize identification of ES cells that incorporated all of these elements, we inserted a second *Frt-Neo-Frt* cassette. Insert the selection cassette within a nearby intron to avoid altering the desired protein sequence after induction (Fig. 1b). This approach allows a single targeting in ES cells for integration of all elements rather than sequential targeting.

4. Plan the homology arms to direct homologous recombination in ES cells. Regions of exact homology between the knock-in vector and the endogenous locus will direct homologous recombination in ES cells. To promote crossover events outside of the engineered elements, use homology arms approximately 4 kb on one side and 2 kb on the other side. For the *H3f3a* targeting vector, short and long homology arms (2 and 4.6 kb, respectively) facilitate homologous recombination of the final 17.5 kb engineered locus (Fig. 1b, c).

5. Plan the strategy to identify correctly targeted ES cells. Southern blot analysis remains the gold standard to assess accurate homologous recombination in ES cells. To discriminate between homologous recombination and random integration, a PCR approach must span from the endogenous locus outside of the knock-in targeting construct and connect with the alterations introduced into the locus. There is no positive control for this event until homologous recombination occurs, and spurious PCR results often give misleading false positives or false negatives. To plan the Southern blot approach, consider overall restriction enzyme digest map of the endogenous locus with additional sites added by engineered elements. Select restriction enzyme sites that reside within and outside the engineered locus and are efficient in cutting genomic DNA. Design probes 500–1000 bp residing outside the targeting construct to provide adequate resolution to discriminate bands from targeted and wild-type alleles (*see* **Note 2**). In general, a range of bands of 5–12 kb with at least a 20% difference between wild-type and targeted fragments is ideal. Test probes on wild-type DNA to verify clear results before proceeding. For the *H3f3a* design, we identified *EcoRI*/ *BamHI* double digest suitable to assess DNA banding patterns

with 5′ probe and *SacI* to assess DNA banding patterns with 3′ probe (Fig. 1a, b).

6. Plan a negative selection element in the knock-in construct. The final targeting vector used in this example also includes diphtheria toxin (DT) outside of the regions of homology to select against ES cells with random integration. The *H3f3a* final knock-in vector confers DT negative selection plus puro and neo double positive selection (Fig. 1c).

7. Identify a unique restriction site to linearize the knock-in construct. The final targeting vector should contain a unique restriction enzyme cut site in the vector backbone that can be used to linearize the construct without disrupting the targeted locus or any engineered elements. If such a site does not exist, you can engineer it into the vector before construction begins. We used *NotI* to linearize the final *H3f3a* knock-in vector (Fig. 1c).

3.2 Retrieve the Genomic Region of Interest from a BAC Clone

This step uses recombineering to isolate the desired genomic region of your gene of interest from the much larger BAC clone and insert into the retrieval vector.

3.2.1 Make Electrocompetent SW102 Bacteria

1. Begin starter culture of SW102 bacteria: 5 µL/5 mL LB with tetracycline (15 µg/mL). Incubate/shake at 30 °C overnight approximately 18 h.

2. Transfer 800 µL starter culture to 40 mL LB with tetracycline (15 µg/mL), save remaining culture as a reference for measuring O.D.$_{600}$. Incubate/shake diluted culture at 30 °C until O.D.$_{600}$ is 0.6.

3. Cool cells in ice water for 2 min. Transfer 12.5 mL of culture into individual ice cold 14 mL round-bottom tube in ice water.

4. Spin 5 min at ~2000 × *g* at 4 °C. Discard supernatant, add 1 mL ice cold nuclease-free water. Hand-shake in ice water slurry until pellet is resuspended. Add 9 mL ice cold nuclease-free water. Invert tubes two times.

5. Repeat **step 4** above.

6. Spin 5 min at ~2000 × *g* at 4 °C. Decant supernatant, be careful not to lose pellet and resuspend by hand-shaking in residual approximately 200 µL nuclease-free water. These are electrocompetent cells, ready for electroporation.

7. If electrocompetent cells are to be saved for later use, replace ice cold water with ice cold 10% glycerol in 3–6 above and store final 200 µL of cells at −80 °C.

<table>
<tr><td>

3.2.2 Electroporate BAC DNA into SW102 Bacteria

</td><td>

1. Transfer 50 µL electrocompetent SW102 cells into ice cold 0.1 cm gap Gene Pulser Electroporation Cuvette. Add 1–10 µL (approximately 1 µg) BAC DNA (124 kb) to cuvette cells. Include cuvettes for two additional conditions: a no DNA negative control, and a known circular plasmid with appropriate antibiotic selection positive control. Electroporate with the following parameters: 1750 V, 200 Ω, 25 µF. Monitor time constant and voltage output (usually approximately 5 and 1725, respectively).

2. Recover electroporated cells in 350 µL S.O.C. medium and incubate/shake at 30 °C for 1 h.

3. Plate recovered cells onto chloramphenicol (12.5 µg/mL) LB agar plates. Incubate overnight at 30 °C.

4. Select 5–10 chloramphenicol-resistant colonies, grow 5 mL cultures in LB-chloramphenicol (12.5 µg/mL) media overnight at 30 °C.

5. Make multiple 10% DMSO stocks from individual colonies by adding 100 µL of DMSO to 900 µL culture. Store at −80 °C. **These are SW102-BAC bacteria.**

6. Isolate DNA from 5 mL cultures using a modified protocol from QIAGEN Large-Construct Kit (*see* **Note 3**).

</td></tr>
<tr><td>

3.2.3 Construct the Retrieval Vector

</td><td>

1. Design retrieval arms at the 5′ and 3′ ends of the genomic region that you would like to include in your knock-in construct (*see* **Notes 4–6**). For *H3f3a*, we retrieved 16.2 kb encompassing the entire *H3f3a* locus.

2. Using the BAC DNA as a template, PCR amplify 5′ retrieval arm (5′ Ret). Engineer complimentary restriction enzyme sites in your PCR primers for ligating into the main targeting vector, pBR322-DT (*see* **Notes 7** and **8**) and assembling with the 3′ retrieval arm. For *H3f3a*, we engineered the 5′ retrieval arm to contain flanking *XhoI* and *HindIII* sites (Fig. 2a, b). TA-subclone the PCR fragment into pCR2.1-TOPO, transform into TOP10 cells, and verify the sequence.

3. Using the BAC DNA as a template, PCR amplify 3′ retrieval arm (3′ Ret). Again, engineer complimentary restriction enzyme sites in your PCR primers for ligating into the main targeting vector, pBR322-DT, and assembling with the 5′ retrieval arm. For *H3f3a*, we used flanking *HindIII/SpeI* and *ClaI* sites (Fig. 2a, b). TA-subclone into pCR2.1-TOPO, transform into TOP10 cells and verify the sequence.

4. Subclone the 5′ and 3′ retrieval arms into pBR322-DT. Linearize pBR322-DT with *XhoI* and *ClaI*, isolate 5′ Ret with *XhoI* and *HindIII*, and 3′ Ret with *HindIII* and *ClaI*. Perform a triple ligation to clone 5′ and 3′ retrieval arms into pBR322-

</td></tr>
</table>

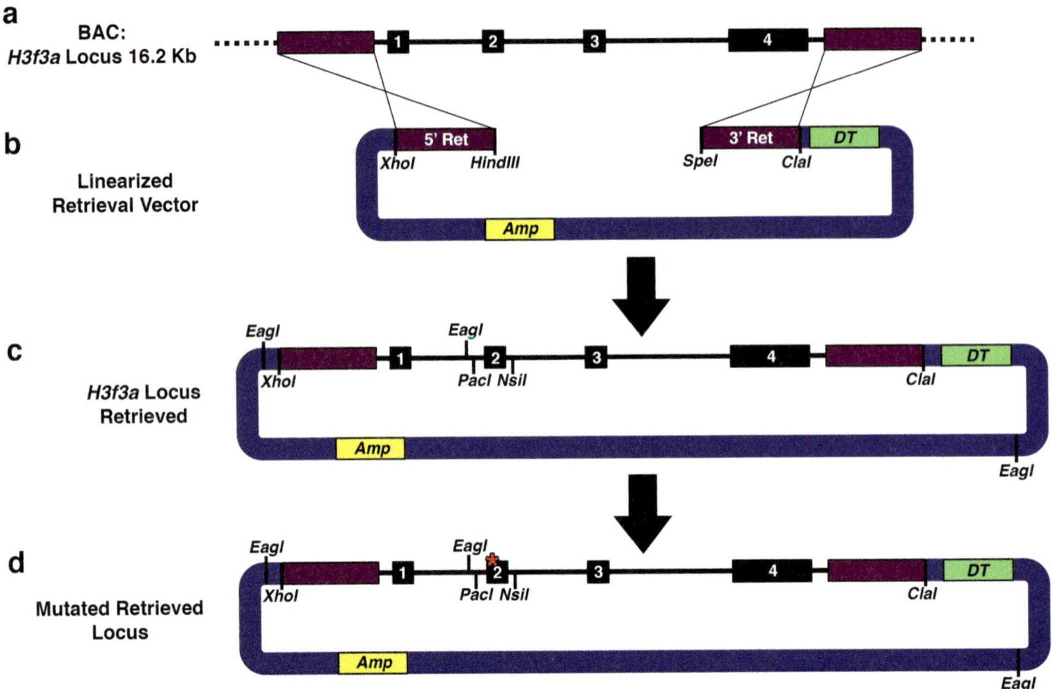

Fig. 2 *H3f3a* Locus Retrieval. (**a**) Schematic of desired 16.2 kb *H3f3a* locus including exons 1–4 within a 124 kb BAC clone. Dotted black line indicates genomic regions outside of region to be retrieved. (**b**) Retrieval vector with specific locus homology retrieval arms (5′ Ret and 3′ Ret), linearized with *HindIII* and *SpeI* to facilitate gap repair homologous recombination. pBR322-DT Vector (blue) is equipped to express Diphtheria toxin (*DT*) for negative selection in mammalian cells and ampicillin resistance (*Amp*) in bacterial cells. (**c**) Successfully retrieved locus. *EagI* sites mark diagnostic digest points to verify vector fidelity. *PacI* and *NsiI* sites flank exon 2 facilitating mutant exon 2 replacement. (**d**) Retrieved locus with successfully replaced mutated exon 2 (red asterisk). If mutation in the gene of interest is located distally to the *loxP*-STOP-*loxP* cassette, it would be introduced by recombineering as a distal element in **Targeting 2** rather than at this early stage

DT. Transform into TOP10 bacteria, extract DNA, and verify by restriction digest. **This is the final retrieval vector** (Fig. 2b).

3.2.4 Retrieve the Genomic Region of Interest into pBR322-DT

1. Linearize retrieval vector by restriction enzyme digest to cut at the junction between the 5′ and 3′ retrieval arms (Fig. 2b). *HindIII* and *SpeI* were used for *H3f3a* retrieval. Purify by gel extraction (QIAquick Gel Extraction Kit).

2. For gap repair-mediated homologous recombination, begin starter culture of SW102-BAC bacteria prepared in Subheading 3.2.2, **step 5**: 5 μL/5 mL LB with chloramphenicol (25 μg/ mL). Incubate/shake at 30 °C overnight approximately 18 h.

3. Transfer 800 μL starter culture to 40 mL LB with chloramphenicol (25 μg/mL), save remaining culture as a reference for

measuring O.D.$_{600}$. Incubate/shake diluted culture at 30°C until O.D.$_{600}$ is 0.6.

4. Separate 12.5 mL of culture into a 50 mL conical tube, keep remainder at 30 °C (to be used as a no heat-induced gap repair negative control that should result in no colonies after selection). Shake 50 mL conical tube in 42 °C water bath for 15 min to heat-shock the bacteria.

5. Shake heat-shocked bacteria in ice water slurry for 2 min. Do the same with 10 mLs of negative control bacteria that were not heat shocked. Incubate each culture in ice water for 5 min. Transfer 10 mL of each culture into individual pre-chilled 14 mL round-bottom tubes in ice water slurry.

6. Spin 5 min at ~2000 × g at 4 °C. Discard supernatant, add 1 mL ice cold nuclease-free water. Hand-shake in ice water slurry until pellet is resuspended. Add 9 mL ice cold nuclease-free water. Invert tubes two times.

7. Repeat **step 6** above.

8. Spin 5 min at ~2000 × g at 4 °C. Decant supernatant, be careful not to lose pellet and resuspend by hand-shaking in residual approximately 200 µL nuclease-free water.

9. For each culture, transfer 50 µL cells into ice cold 0.1 cm gap gene Pulser Electroporation cuvette. Add 1–10 µL (approximately 20 ng) linearized retrieval vector to each cuvette. Include two additional controls for each culture: a no DNA negative control, and a known circular plasmid with appropriate antibiotic selection positive control. Electroporate with the following parameters: 1750 V, 200 Ω, 25 µF. Monitor time constant and voltage output (usually approximately 5 and 1725, respectively).

10. Recover electroporated cells in 350 µL S.O.C. medium and incubate/shake at 30 °C for 1 h.

11. Plate recovered cells onto LB-ampicillin (100 µg/mL) agar plates. Incubate overnight at 30 °C.

12. Select 5–10 colonies, grow 5 mL cultures in LB-ampicillin (100 µg/mL) media overnight at 30 °C, isolate DNA from a portion of each culture (*see* **Note 3**) **This is the retrieved locus vector** (Fig. 2b).

13. Perform diagnostic *Eag*I restriction enzyme digest to verify correct banding pattern for the retrieved locus vector (Figs. 2c and 7). Make 10% DMSO stocks of validated clones, store at −80 °C.

3.3 Engineer Proximal Point Mutations

In this example, we introduced a K27M mutation into *H3f3a* exon 2.

1. Using the BAC DNA as a template, PCR amplify *H3f3a* exon 2 and flanking *PacI* and *NsiI* restriction enzyme sites (Fig. 2c). TA-subclone into pCR2.1-TOPO and verify sequence.

2. Design appropriate primers and use exon 2 pCR2.1-TOPO subclone as a template and perform site-directed mutagenesis PCR (QuickChange II SDM Kit) to incorporate desired mutation.

3. Verify sequence of mutated exon 2.

4. Linearize retrieved locus vector and isolate mutated exon 2 fragment with *PacI*/*NsiI* double digest. Since the retrieved locus vector is >10 kb, gel purify pieces with QIAEX II Gel Extraction Kit.

5. Ligate mutated exon 2 fragment into retrieved locus vector. Purify half of the ligations with QIAEX II gel extraction kit to maximize electroporation efficiency in next step.

6. Electroporate MegaX DH10B T1^R Electrocompetent *E. coli* cells with the following parameters: 2000 V, 200 Ω, 25 μF.

7. Recover electroporated cells in 350 μL S.O.C. medium and incubate/shake at 30 °C for 1 h.

8. Plate recovered cells onto LB-ampicillin (100 μg/mL) agar plates. Incubate overnight at 30 °C.

9. Select 5–10 colonies, grow 5 mL cultures in LB-ampicillin (100 μg/mL) media overnight at 30 °C.

10. Isolate DNA from a portion of each culture (*see* **Note 3**).

11. Confirm *EagI* digest banding pattern (Figs. 2d and 7) and verify mutated exon 2 sequence (Fig. 2d) with primers flanking exon 2. Make 10% DMSO stocks of validated clones.

12. Electroporate 1–10 μL (approximately 20 ng) of this exon 2 mutated retrieved locus vector back into SW102 bacteria following steps in Subheading 3.2.1 to make them electrocompetent and steps in Subheading 3.2.2, **steps 1–3** for electroporation selecting with LB-ampicillin (100 μg/mL).

13. Select 5–10 colonies, grow 5 mL cultures in LB-ampicillin (100 μg/mL) media overnight at 30 °C.

14. Isolate DNA from a portion of each culture (*see* **Note 3**). **This is the mutated retrieved locus vector** (Fig. 2d).

15. Re-verify *EagI* digest banding pattern (Figs. 2d and 7). Make 10% DMSO stocks of validated clones and use these SW102 bacteria containing the mutated retrieved locus for subsequent engineering.

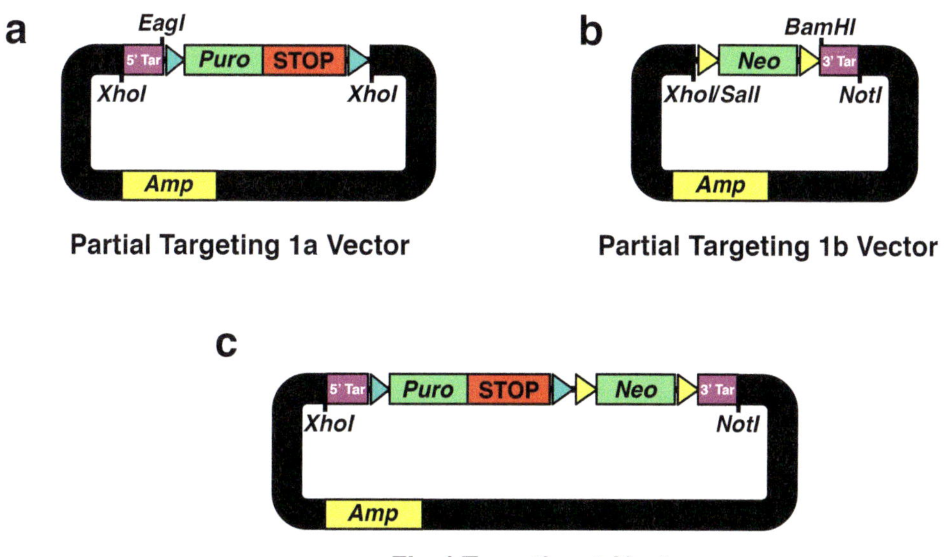

Fig. 3 Assembling Targeting 1 Vector For *loxP*-STOP-*loxP* Insertion. (**a**) Partial targeting 1a vector containing subcloned *XhoI*-flanked *H3f3a* intron 1 5′ homology targeting arm (5′ Tar) and *loxP*-STOP-*loxP* (STOP) cassette equipped to express puromycin (*Puro*) in mammalian cells. *EagI* is the designed diagnostic subcloning remnant restriction enzyme site. (**b**) Partial targeting 1b vector containing subcloned *XhoI/SalI* and *NotI*-flanked *Frt-Neo-Frt* (*Neo*) cassette equipped to express neomycin in bacterial and mammalian cells, and *H3f3a* intron 1 3′ targeting homology arm (3′ Tar). *BamHI* is a subcloning remnant restriction enzyme site. (**c**) Final targeting 1 vector. *loxP* sites (cyan arrowheads), *Frt* sites (yellow arrowheads), ampicillin resistance in bacterial cells (*Amp*)

3.4 Targeting 1: Inserting the loxP-STOP-loxP Cassette

To insert the *loxP*-STOP-*loxP* cassette, build a construct with 5′ targeting and 3′ targeting arms to direct integration by homologous recombination at the desired location in the mutated retrieved locus. Include a *Frt-Neo-Frt* cassette, which confers kanamycin resistance to bacteria for selecting desired clones.

3.4.1 Constructing the loxP-STOP-loxP Targeting 1 Vector

1. Using the BAC DNA as a template, PCR amplify 5′ targeting arm (*see* **Notes 9** and **10**) (5′ Tar, Fig. 3a) to contain flanking *XhoI* and *EagI* sites, and TA-subclone into pCR2.1-TOPO, transform into TOP10 cells and verify sequence.

2. Isolate 5′ Tar by double digest with *XhoI/EagI*, *loxP*-STOP-*loxP* cassette by *NotI/XhoI* from pLox-Stop-Lox-TOPO, and linearize pBluescript KS(+) with *XhoI*. Gel purify these 3 fragments and perform a triple ligation to assemble a partial targeting 1a vector backbone (Fig. 3a). This ligation eradicates *NotI* site between 5′ Tar and *loxP*-STOP-*loxP* cassette, and leaves the *EagI* restriction enzyme site for diagnostic size verification of the assembled targeting vector at this stage (*see* **Note 10**).

3. Transform into chemically competent Stbl3 cells (*see* **Note 11**), incubate overnight at 30 °C and select with ampicillin (100 μg/ mL). Select 5–10 clones and verify correct ligation using *EagI* restriction site banding pattern.

4. Using the BAC DNA as a template, PCR amplify 3′ targeting arm (*see* **Note 9**), (3′ Tar, Fig. 3b) to contain flanking *BamHI* and *NotI* sites, and TA-subclone into pCR2.1-TOPO, transform into TOP10 cells and verify sequence.

5. Isolate 3′ Tar fragment by double digest with *BamHI/NotI*, *Frt-Neo-Frt* cassette from PL451-no *loxP* plasmid by *XhoI/ BamHI* and linearize pBluescript KS(+) backbone with *XhoI/ NotI*. Gel purify these 3 fragments and perform a triple ligation to assemble a partial targeting 1b vector backbone (Fig. 3b).

6. Transform into chemically competent Stbl3 cells (*see* **Note 11**). Incubate overnight at 30 °C and select with kanamycin (50 μg/ mL). Select 5–10 clones and verify correct ligation using *EagI* or *KpnI* restriction site banding pattern.

7. Isolate partial targeting 1a fragment by *XhoI* digest, partial targeting 1b fragment by *SalI/NotI* double digest, and linearize pBluescript KS(+) backbone with *SalI/NotI*. Gel purify these 3 fragments and perform a triple ligation to assemble complete targeting 1 vector (Fig. 3c). This ligation eradicates *XhoI/SalI* sites between partial targeting 1a targeting 1b fragments).

8. Transform into chemically competent Stbl3 cells (*see* **Note 11**) and select with kanamycin (50 μg/mL). Select 5–10 clones and verify correct ligation using separate *XhoI/NotI*, *BamHI* or *EcoRI* restriction site banding pattern. **This is the targeting 1 vector** to insert the *loxP*-STOP-*loxP* cassette (Fig. 3c).

3.4.2 Insert loxP-STOP-loxP/Neo Selection Cassette into Mutated Retrieved Locus Vector

1. Isolate targeting 1 cassette from targeting 1 vector with *XhoI/ NotI* double digest (Fig. 4a). Purify by gel extraction using QIEX II Gel Extraction Kit.

2. Using SW102 bacteria containing mutated retrieved locus from Subheading 3.3, follow Subheading 3.2.4, **steps 2–12** with LB-ampicillin (100 μg/mL) selection prior to electroporation and LB-kanamycin (50 μg/mL) selection after electroporating isolated targeting cassette 1.

3. Perform diagnostic *EagI* restriction enzyme digest to verify correct banding pattern for targeting 1 (Figs. 4b and 7).

4. Select one correct targeting 1 clone DNA from Subheading 3.4.2, **step 3** above, electroporate into MegaX DH10B T1^R Electrocompetent *E. coli* cells following Subheading 3.3, **steps 6–10** above selecting with LB-kanamycin (50 μg/mL).

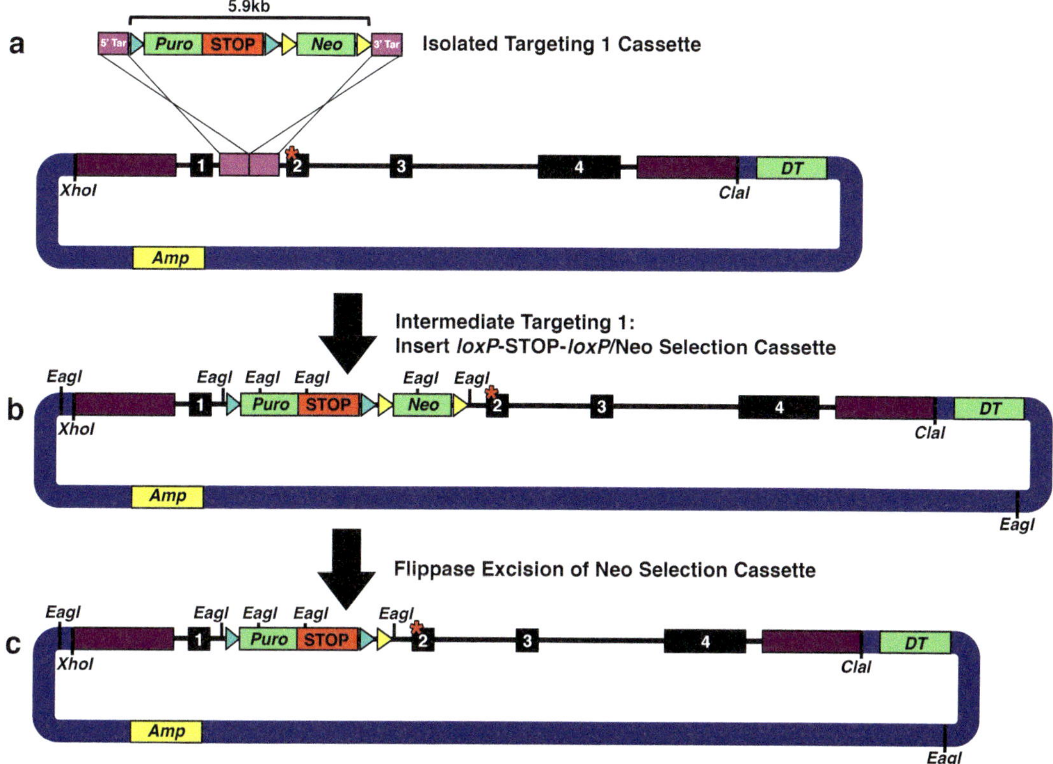

Fig. 4 Targeting 1: Inserting *loxP*-STOP-*loxP*. (**a**) Isolated targeting 1 cassette excised with *Xhol* and *Notl* to facilitate gap repair homologous recombination. (**b**) Intermediate step of targeting 1: Successful insertion of targeting 1 *loxP*-STOP-*loxP*/*Neo* selection cassette (5.9 kb) into *H3f3a* intron 1. *Eagl* sites mark diagnostic digest points to verify vector fidelity. (**c**) Successful Flippase excision of *Neo* selection cassette leaving behind a remnant *Frt* site and stable *loxP*-STOP-*loxP*. *Eagl* sites mark diagnostic digest points to verify vector fidelity

Grow cultures and isolate DNA from a portion of each culture (*see* **Note 3**).

5. Re-verify *EagI* digest banding pattern (Figs. 4b and 7). Make 10% DMSO stocks of validated clones. **This is the intermediate step of targeting 1** (Fig. 4b).

3.4.3 Flippase-Mediated in vitro Removal of Targeting 1 Neo Selection Cassette

1. Begin starter culture of SW105 bacteria 5 μL/5 mL LB with no antibiotic. Incubate/shake at 30 °C overnight approximately 15 h.

2. Transfer 800 μL starter culture to 40 mL LB with no antibiotic, make 1 duplicate dilution and save remaining culture as a reference for measuring O.D.$_{600}$. Incubate/shake diluted cultures at 30 °C until O.D.$_{600}$ is 0.4.

3. To 1 duplicate culture dilution, add 400 μL 10% L-(+)-Arabinose to induce expression of Flippase.

4. Incubate/shake both diluted cultures at 30 °C for 1 h.

5. Make cells electrocompetent using conditions in Subheading 3.2.1, **steps 3–6**.

6. For each culture, transfer 50 μL cells into 2 separate ice-cold 0.1 cm gap gene Pulser Electroporation cuvette (sets 1a and 1b). Add 1–10 μL (approximately 20 ng) intermediate targeting 1 *loxP*-STOP-*loxP* + *Neo* vector DNA produced in Subheading 3.4.2 to each cuvette. For each culture, transfer another 50 μL cells into 2 separate ice cold 0.1 cm gap gene Pulser electroporation cuvette to serve as no DNA controls (sets 2a and 2b). For each culture, also prepare 2 cuvettes to electroporate MegaX DH10B T1^R Electrocompetent *E. coli* cells with a 1:100 dilution of targeting 1 clone DNA (sets 3a and 3b) to serve as electroporation controls. Include one additional electroporation for each culture: a known circular plasmid with appropriate antibiotic selection positive control. Electroporate with the following parameters: 1750 Volts, 200 Ω, 25 μF. Monitor time constant and voltage output (usually approximately 5 and 1725, respectively).

7. Recover electroporated cells in 350 μL S.O.C. medium and incubate/shake at 30 °C for 1 h.

8. Plate half of recovered cells for each set onto LB-ampicillin (100 μg/mL) and the other half on kanamycin (50 μg/mL) agar plates. Incubate overnight at 30 °C. Colony growth on ampicillin plates and lack of growth on kanamycin plates after arabinose treatment indicates efficiency of Flippase-mediated excision of *Neo* cassette. Colony growth on kanamycin plates without arabinose treatment indicates the specificity of the arabinose-induced Flippase expression in the SW105 bacteria.

9. Select 5–10 colonies, grow 5 mL cultures in LB-ampicillin (100 μg/mL) media overnight at 30 °C, isolate DNA from a portion of each culture (*see* **Note 3**).

10. Perform diagnostic *EagI* restriction enzyme digest to verify correct banding pattern (Figs. 4c and 7). Make 10% DMSO stocks of validated clones.

3.4.4 Transfer Mutant Locus + loxP-STOP-loxP Vector into SW102 Cells

1. Make new electrocompetent SW102 cells as in Subheading 3.2.1.

2. Transfer in duplicate 50 μL cells into ice cold 0.1 cm gap gene Pulser electroporation cuvette. Add 1–10 μL (approximately 20 ng) mutant locus + targeting 1 vector (from Subheading 3.4.3) to cuvette cells. Include duplicate cuvettes for two additional conditions: a no DNA negative control, and a known circular plasmid with appropriate antibiotic selection positive control. Electroporate with the following parameters: 1750 V, 200 Ω, 25 μF. Monitor time constant and voltage output (usually approximately 5 and 1725, respectively).

3. Recover electroporated cells in 350 μL S.O.C. medium and incubate/shake at 30 °C for 1 h.

4. Plate recovered cells onto LB-ampicillin (100 μg/mL) or kanamycin (50 μg/mL) agar plates. Incubate overnight at 30 °C.

5. Colony growth on kanamycin plates indicates contamination of un-excised *Neo* cassette. Select 5–10 ampicillin-resistant colonies, grow 5 mL cultures in LB-ampicillin (100 μg/mL) media overnight at 30 °C, isolate DNA from a portion of each culture (*see* **Note 3**).

6. Reconfirm diagnostic *EagI* restriction enzyme digest banding pattern (Figs. 4c and 7). Make 10% DMSO stocks of validated clones. **This completes targeting 1: *loxP*-STOP-*loxP* inserted into mutant retrieved locus** (Fig. 4c). If there were no additional distal genetic elements to engineer, this would be the finished knock-in targeting construct. Use these SW102 bacteria containing the mutated retrieved locus + *loxP*-STOP-*loxP* vector for subsequent engineering.

3.5 Targeting 2: Distal Genetic Element and Neo Selection Cassette

To insert an additional genetic element, build a construct with 5′ targeting and 3′ targeting arms to direct integration by homologous recombination at the desired location. To increase the frequency of recovering ES clones that have incorporated all of the engineered events, include a *FrtF3-Neo-FrtF3* cassette so that double selection can be used. The *FrtF3* sites will not recombine with the residual *Frt* site adjacent to the *loxP*-STOP-*loxP* cassette (*see* **Note 12**).

3.5.1 Constructing Targeting 2 Distal Element Insertion Vector

This vector is designed to incorporate an additional genetic element, which could include alterations to the coding sequence of the target gene, or insertions of epitope tags or GFP fusions. As before, targeting arms include approximately 300 bp of endogenous sequence flanking any engineered sequence. In the *H3f3a* example, we inserted an in frame epitope tag immediately upstream of the endogenous exon 4 translation STOP codon. Since this epitope will be approximately 8.5 kb away from the exon 2 mutation, a *FrtF3*-flanked *Neo* cassette was included upstream of exon 4 to facilitate double positive selection when isolating ES clones.

1. Using the BAC DNA as a template, PCR amplify 5′ targeting arm (5′ Tar) to contain flanking *HindIII* and *EcoRI* restriction enzyme sites (*see* **Note 9**). TA-subclone into pCR2.1-TOPO to assemble partial targeting 2a vector (Fig. 5a). Verify sequence.

2. Using the BAC DNA as a template, PCR amplify 3′ targeting arm (3′ Tar), to contain flanking *BglII/SacI* and *SacII* restriction enzyme sites (*see* **Note 9**). For *H3f3a*, this arm includes the endogenous STOP codon, and the engineered *SacI* restriction enzyme site serves as a new DNA cut site for Southern blot

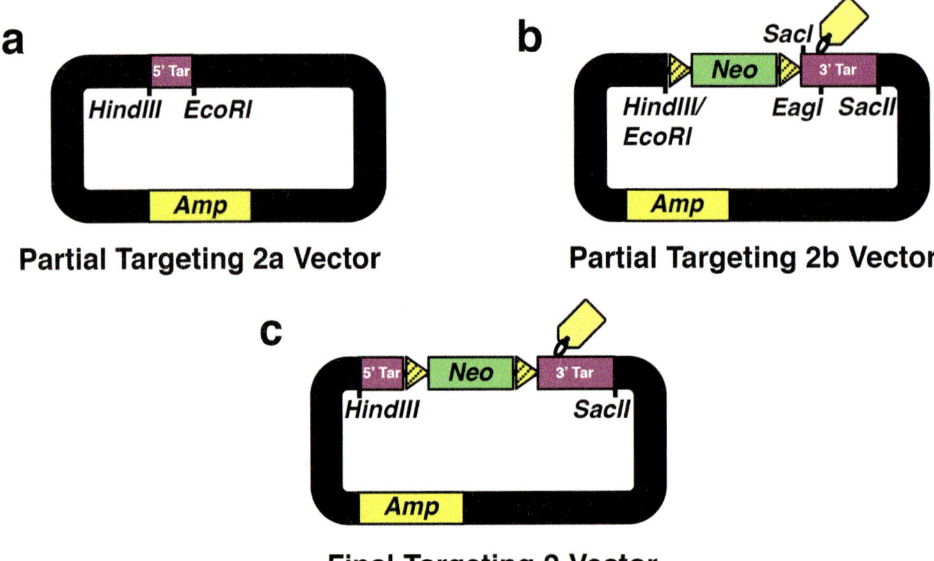

Fig. 5 Assembling Targeting 2 Vector For Distal Element Insertion. (**a**) Partial targeting 2a vector containing subcloned *HindIII and EcoRI*-flanked *H3f3a* intron 3 5′ targeting homology arm (5′ Tar). (**b**) Partial targeting 2b vector containing subcloned *HindIII/EcoRI* and *SacII*-flanked *Frt-Neo-Frt* (*Neo*) cassette to express neomycin in bacterial and mammalian cells and *H3f3a* intron 3/exon 4 3′ targeting homology arm (3′ Tar). *SacI* is the designed subcloning remnant restriction enzyme site for Southern blot verification. (**c**) Final targeting 2 vector. Mutated *Frt* sites (striped yellow). 3′ Tar is engineered to contain the diagnostic linker *EagI* restriction enzyme site and epitope tag sequence

verification (Fig. 5b). TA-subclone into pCR2.1-TOPO and verify sequence.

3. Add the epitope tag to the 3′ targeting arm using the 3′ Tar subclone as a template for splice overlap extension PCR [13]. The epitope was designed with a linker sequence containing an *EagI* restriction enzyme site for diagnostic size verification of the assembled targeting vector at this stage (Fig. 5b).

4. Isolate 3′ Tar + epitope insert by double digest with *BglII/SacII*, *FrtF3-Neo-FrtF3* cassette from p*FrtF3-Neo-FrtF3* plasmid by *HindIII/BamHI* and linearize pBluescript KS(+) backbone with *HindIII/SacII*. Gel purify these 3 fragments and perform a triple ligation to assemble a partial targeting 2b vector (Fig. 5b). This ligation eradicates *BamHI* and *BglII* sites between 3′ Tar and *Neo* cassette.

5. Subclone 5′ Tar arm into the partial targeting 2b vector by isolating 5′ Tar insert by *HindIII/EcoRI* double digest and linearize partial targeting 2b vector backbone above by *HindIII/EcoRI*. Gel purify these 2 fragments and ligate to assemble **final targeting 2 vector** (Fig. 5c).

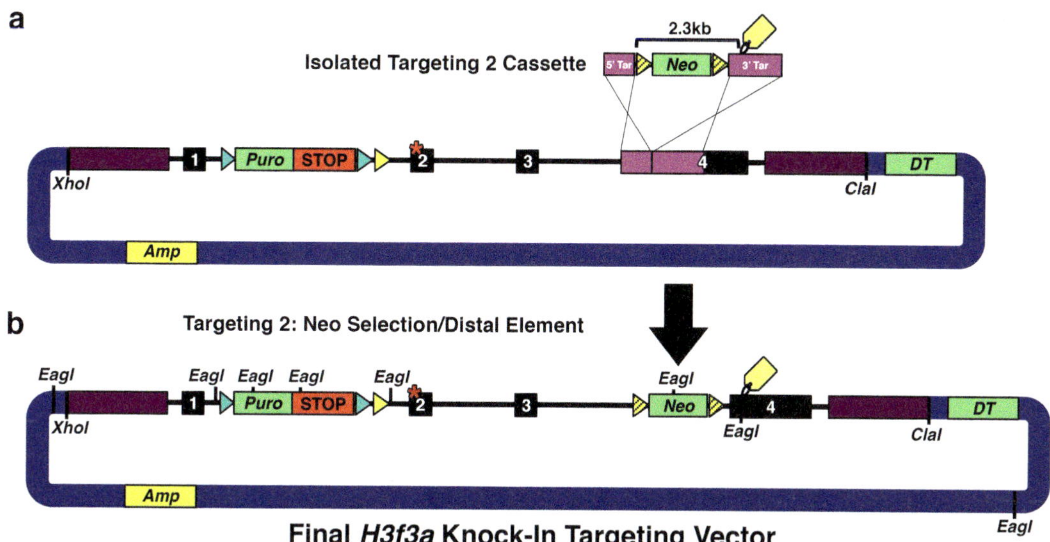

Final *H3f3a* Knock-In Targeting Vector

Fig. 6 Targeting 2: Inserting Distal Element. (**a**) Targeting 2 cassette fragment isolated with *HindIII* and *SacII* to facilitate gap repair homologous recombination. (**b**) Final *H3f3a* knock-in targeting vector generated by successful insertion of targeting 2 *Neo* selection/distal element cassette (2.3 kb) into *H3f3a* intron 3/exon 4. *EagI* sites mark diagnostic digest points to verify final gene targeting vector fidelity

6. Transform into chemically competent Stbl3 cells (*see* **Note 11**) and select with kanamycin (50 μg/mL). Select 5–10 clones and verify correct ligation using diagnostic restriction enzyme cutting patterns.

3.5.2 Insert Distal Genetic Element and Neo Selection Cassette into Mutant Locus + loxP-STOP-loxP Targeting 1 Vector

1. Isolate targeting 2 cassette from final targeting 2 vector by *HindIII/SacII* double digest (Fig. 6a). Purify by gel extraction (QIAquick gel extraction kit).

2. Using SW102 bacteria containing mutant retrieved locus + *loxP*-STOP-*loxP* vector (from Subheading 3.4.4), follow Subheading 3.2.4, **steps 2–12** with LB-ampicillin (100 μg/mL) selection prior to electroporation and LB-kanamycin (50 μg/mL) selection after electroporating isolated targeting cassette 2.

3. Perform diagnostic *EagI* restriction enzyme digest to verify correct banding pattern (Figs. 6b and 7). Make 10% DMSO stocks of validated clones.

4. Select one correct clone DNA from Subheading 3.5.2, **step 3** above.

5. Electroporate MegaX DH10B T1^R Electrocompetent *E.coli* cells with up to 100 ng DNA using the following parameters: 2000 V, 200 Ω, 25 μF. Also electroporate a no DNA negative control and a known circular plasmid with appropriate antibiotic selection positive control.

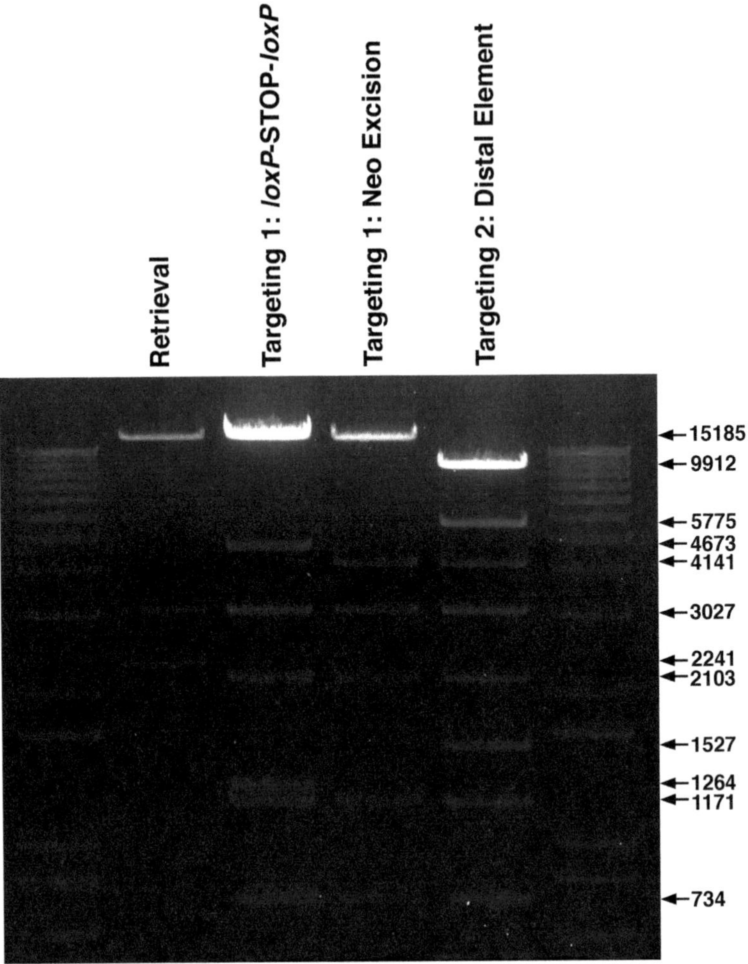

Fig. 7 Verifying *H3f3a* Knock-In Targeting Vector Construction. *Eag*I restriction enzyme digests to verify correct assembly for each step of *H3f3a* knock-in targeting vector construction

6. Recover electroporated cells in 350 μL S.O.C. medium and incubate/shake at 30 °C for 1 h.

7. Plate recovered cells onto LB-kanamycin (50 μg/mL) agar plates. Incubate overnight at 30 °C.

8. Select 2–3 colonies, grow 5 mL cultures in LB-kanamycin (50 μg/mL) media overnight at 30 °C, isolate DNA from a portion of each culture (*see* **Note 3**).

9. Re-verify *Eag*I digest banding pattern (Figs. 6b and 7). Make 10% DMSO stocks of validated clone.

10. **This is the final *H3f3a* knock-in targeting vector** (Fig. 6b).

3.6   Homologous Recombination into Mouse ES Cells

3.6.1   Prepare Linearized Construct for ES Cell Electroporation

1. Linearize 40 µg of final knock-in targeting vector with unique restriction site that cuts in the vector backbone. *NotI* was used for this *H3f3a* example.

2. Bring reaction volume to 350 µL with nuclease-free water.

3. Extract with 350 µL P:C:I.

4. Spin 5 min at ~16,000 × *g*. Transfer aqueous (top) phase to a fresh tube (~350 µL).

5. Extract with equal volume chloroform (~350 µL).

6. Spin 5 min at ~16,000 × *g*. Transfer aqueous (top) phase to a fresh tube (~350 µL).

7. Add 1/10 volume 3 M sodium acetate (~30 µL), mix gently.

8. Add 660 µL ethanol, mix gently. DNA cloud becomes visible.

9. Spin at ~16,000 × *g*, 4 °C for 15 min.

10. Remove supernatant and wash pellet with 1 mL ice cold 70% ethanol.

11. Spin at ~16,000 × *g*, 4 °C for 10 min.

12. Repeat 70% ethanol wash and spin.

13. Remove supernatant and air dry pellet.

14. Resuspend in 10 µL LoTE.

3.6.2   Electroporate ES Cells with Linearized Knock-In Targeting Construct

For guidelines on propagating ES cells, including all media and solutions, conditions to generate irradiated MEF feeder cells, picking ES clones, and DNA extraction from clones, please refer to [10].

1. Grow ES cells on a monolayer of irradiated DR4 MEFs.

2. When ES cells reach 80% confluence, trypsinize in 2 mL, add 8 mL media.

3. Spin 5 min at ~180 × *g*.

4. Wash pellet gently with 10 mL PBS.

5. Spin 5 min at ~180 × *g*.

6. Resuspend pellet gently in 1.6 mL ice cold PBS.

7. Divide dilution into two 800 µL aliquots and add to ice cold 0.4 cm gap Gene Pulser Electroporation cuvette.

8. To each cuvette, add 20 µg linearized targeting vector DNA from Subheading 3.6.1.

9. Electroporate with the following parameters: 250 V, 500 µF. Monitor time constant and voltage output (usually approximately 7 and 250, respectively). Add each electroporation to 30 mL pre-warmed media.

10. Split each electroporation onto three 10 cm^2 plates seeded with irradiated MEFs.

11. Incubate cells overnight.

12. Replace media with double-positive selection media (2 μg/mL Puromycin, 100 μg/mL G418).

13. Replace double positive selection media every day for 4 days. Small colonies become visible.

14. Replace double positive selection media with G418 selection media (100 μg/mL G418) every day until colonies become large enough to pick (approximately 1 week after electroporation).

15. Pick clones, expand to create duplicate clone sets. Freeze viable cells for one set and use duplicate set for DNA extraction and Southern blot analysis to identify correctly targeted clones (*see* **Note 13**).

16. Correctly targeted ES clones can be used for blastocyst injection and subsequent breeding to create knock-in mice, following procedures outlined in detail in [10] (*see* **Notes 14 and 15**).

4   Notes

1. Obtain mouse BAC clones generated from the same mouse strain (e.g., 129) as the mouse ES cells to be used for targeting. This will aid in proper homologous recombination of the final targeting vector in ES cells.

2. The design for homologous recombination events and Southern blot probes should avoid sequences with repetitive elements when possible to help ensure specificity. RepeatMasker on UCSC genome browser (http://genome.ucsc.edu/), as well as BLAST searches can help identify repetitive elements.

3. Modified protocol for isolating large construct DNA. Pellet 4 mL of culture, resuspend in 250 μL chilled QIAGEN P1 with RNAse A. Add 250 μL QIAGEN P2, mix gently by inverting five times, incubate at room temperature 5 min. Add 350 μL chilled QIAGEN P3, mix gently by inverting five times, incubate on ice for 10 min. Spin at ~16,000 × g for 10 min at 4 °C. Gently transfer supernatant to a fresh tube. Add 750 μL isopropanol, mix gently by inverting five times, incubate at room temperature 10 min. Spin at ~16,000 × g at 4 °C, gently remove supernatant without disturbing DNA pellet. Wash pellet with 1 mL ice cold 70% ethanol. Spin at ~16,000 × g at 4 °C, gently remove supernatant without disturbing DNA pellet. Wash pellet with 1 mL ice cold 70% ethanol. Spin at ~16,000 × g at 4 °C, gently remove supernatant without disturbing DNA pellet. Air dry pellet. Resuspend in 10 μL LoTE.

4. The fragment size to be retrieved will vary depending on the locus and design. Retrieve the region that you need to engineer as well as 5′ and 3′ homology arms.

5. Retrieval arms should ideally be approximately 500 bp to help enrich for correctly targeted fragments [9].

6. Consider designing retrieval vector to linearize with two incompatible restriction enzymes to avoid self-ligation during retrieval.

7. We use pBR322-DT as the vector backbone for the knock-in targeting vector because it is more amenable to cloning and growth of large DNA inserts compared to newer generation cloning vectors.

8. Plan the full strategy before beginning. The order of introducing different elements may vary depending on the specific locus and design. For example, because Subheading 3.3 required digestion of a *PacI* site in the *H3f3a* locus, we eliminated the *PacI* site in the vector backbone of pBR322-DT by Klenow fill-in and blunt-end ligation before Subheading 3.2.

9. Targeting arms should be approximately 300 bp to help enrich for correctly targeted fragments [9].

10. Subcloning strategy to build the *loxP*-STOP-*loxP* targeting vector using restriction enzymes should be carefully considered to avoid PCR amplifying the *loxP*-STOP-*loxP* cassette sequence due to its tetrameric SV40 polyA tandem array (3.5 kb) [12]. This array can undergo spontaneous recombination resulting in loss of one or more of the SV40 polyA elements (860 bp each), therefore it is very important to monitor the proper size of this array with diagnostic restriction enzyme digest at each step of cloning.

11. Use chemically competent Stbl3 cells combined with selected growth at 30 °C when cloning unstable vectors containing repeat sequences such as the *loxP*-STOP-*loxP* and *Frt-Neo-Frt* cassettes.

12. For distal element insertion in **Targeting 2** (Subheading 3.5), we used a *Frt* F3 spacer mutant that was shown to maintain self-recognition, but lack cross-recombination potential with unmutated *Frt* sites [11]. This prevents Flippase-mediated recombination between *FrtF3* sites flanking the *Neo* selectable marker in the distal element and a residual *Frt* site that remains in intron 1 after **Targeting 1**.

13. Targeting efficiency will vary depending on locus targeted and specific knock-in targeting construct.

14. The positive selection *Neo* cassette can be removed by transient ES cell transfection or transduction with a vector expressing Flippase, followed by isolation and screening of individual

clones. Alternatively, to avoid additional ES cell manipulation, the resulting mice can be bred with mice engineered to ubiquitously express Flippase (e.g., 129S4/SvJaeSor-*Gt(ROSA)26Sor^{tm1(FLP1)Dym}*/J, Jackson labs #003946) to delete the *Neo* cassette in the germline of all progeny mice.

15. The combination of multiple mutated alleles is generally required for spontaneous mouse brain tumor formation [5, 14, 15]. Depending on the oncogene, knock-in oncogenic mutations would typically need to be combined with additional mutations through breeding with other engineered mice, or by layering on injection-mediated delivery of other mutations.

Acknowledgments

This work was supported in part by NIH grant CA096832 to SJB, and by ALSAC.

References

1. Wiesner SM, Decker SA, Larson JD et al (2009) De novo induction of genetically engineered brain tumors in mice using plasmid DNA. Cancer Res 69:431–439

2. Hambardzumyan D, Amankulor NM, Helmy KY et al (2009) Modeling adult gliomas using RCAS/t-va technology. Transl Oncol 2:89–95

3. Yang H, Wang H, Shivalila CS et al (2013) One-step generation of mice carrying reporter and conditional alleles by CRISPR/Cas-mediated genome engineering. Cell 154:1370–1379

4. Wang H, Yang H, Shivalila CS et al (2013) One-step generation of mice carrying mutations in multiple genes by CRISPR/Cas-mediated genome engineering. Cell 153:910–918

5. Rankin SL, Zhu G, Baker SJ (2012) Review: insights gained from modelling high-grade glioma in the mouse. Neuropathol Appl Neurobiol 38:254–270

6. Kersten K, de Visser KE, van Miltenburg MH et al (2017) Genetically engineered mouse models in oncology research and cancer medicine. EMBO Mol Med 9:137–153

7. Sanchez-Rivera FJ, Jacks T (2015) Applications of the CRISPR-Cas9 system in cancer biology. Nat Rev Cancer 15:387–395

8. Huijbers IJ, Del Bravo J, Bin Ali R et al (2015) Using the GEMM-ESC strategy to study gene function in mouse models. Nat Protoc 10:1755–1785

9. Liu P, Jenkins NA, Copeland NG (2003) A highly efficient recombineering-based method for generating conditional knockout mutations. Genome Res 13:476–484

10. Behringer R, Gertsenstein M, Nagy K et al (2014) Manipulating the mouse embryo: a laboratory manual, Fourth edn. Cold Spring Harbor Laboratory Press, Cold Spring Harbor, New York

11. Turan S, Kuehle J, Schambach A et al (2010) Multiplexing RMCE: versatile extensions of the Flp-recombinase-mediated cassette-exchange technology. J Mol Biol 402:52–69

12. Jackson EL, Willis N, Mercer K et al (2001) Analysis of lung tumor initiation and progression using conditional expression of oncogenic K-ras. Genes Dev 15:3243–3248

13. Ho SN, Hunt HD, Horton RM et al (1989) Site-directed mutagenesis by overlap extension using the polymerase chain reaction. Gene 77:51–59

14. Zhu G, Rankin SL, Larson JD et al (2017) PTEN signaling in the postnatal perivascular progenitor niche drives medulloblastoma formation. Cancer Res 77:123–133

15. Chow LM, Endersby R, Zhu X et al (2011) Cooperativity within and among Pten, p53, and Rb pathways induces high-grade astrocytoma in adult brain. Cancer Cell 19:305–316

Chapter 19

In Vivo Murine Models of Brain Metastasis

Mohini Singh, Neil Savage, and Sheila K. Singh

Abstract

Metastases are the most common tumor type to affect the adult central nervous system. In vivo modeling of brain metastases provides insight into the mechanisms of metastatic development as well as a clinically relevant therapeutic screening platform. Here we describe the development of a novel mouse model of brain metastasis from a primary lung cancer utilizing primary patient samples. These models provide an accurate representation of different stages of the clinical progression of the disease.

Key words Brain metastasis, In vivo model, Mouse model

1 Introduction

Metastases are the most frequent neoplasm to affect the adult central nervous system, occurring ten times more often than that of primary brain tumors. Of the various sites brain metastases can arise from, lung carcinomas predominate, accounting for 40–60% of brain metastasis cases [1]. This is followed by breast cancers with 15–20% of cases, and melanoma with 5–10% of cases [1]. 70% of cancer patients present with corresponding metastases upon diagnosis, and approximately 40% of these patients will develop systemic brain metastases [1, 2]. Left untreated, these patients have an average survival of a mere 1–2 months [1].

The metastatic process involves complex mechanisms and pathways which are tightly regulated. The metastatic cell must first break away from the tumor to invade the surrounding tissue, before intravasating into a nearby blood or lymphatic vessel. Once in the circulation, the cell must overcome several lethal barriers such as sheer forces and the host immune response before it can arrest at a secondary location. Here the cell will extravasate through the blood-brain barrier (BBB) into the new environment to either colonize the tissue or enter a dormant state [3]. Current research suggests that within the primary tumor exists a specific population that are responsible for metastatic development. These cells

Sheila K. Singh and Chitra Venugopal (eds.), *Brain Tumor Stem Cells: Methods and Protocols*, Methods in Molecular Biology, vol. 1869, https://doi.org/10.1007/978-1-4939-8805-1_19, © Springer Science+Business Media, LLC, part of Springer Nature 2019

resemble that of a cancer stem cell (CSC) population with the ability to self-renew and initiate tumor growth; however, they also possess the ability to undergo and survive the metastatic cascade as well as form tumors in a completely different secondary environment [4]. We have termed cells metastasizing to the brain as brain metastasis initiating cells, or BMICs.

The use of in vivo models to interrogate brain metastases is a powerful tool to investigate the intricacies of metastatic development as seen with clinical disease. A variety of spontaneous, induced, and experimental brain metastasis models have been established that allow the interrogation of different stages of the metastatic process. Metastasis can be induced either by genetic engineering, chemical or UV exposure (in the case of melanoma) to alter the expression of oncogenes or tissue specific deletion systems [5, 6]. However, despite allowing examination of the homing properties of metastatic cells, it is unclear as to whether these methods properly reflect disease progression in humans [6].

Conversely, experimental models allow the researcher more control over the model development, such as cell number and the site of metastasis. The type of model can differ with the material injected and the site of implantations. Syngeneic models are developed from injection of cells of the same strain as the animal model, whereas models developed from introduction of human cells are termed xenogeneic [5, 6]. Models derived from cells that are implanted into the original environment they were obtained are termed orthotopic, and when injected into a different tissue or organ are termed heterotropic. Another vital matter for the development of brain metastases is the route of inoculation. The best representation of metastatic development is to inoculate at the primary site of origin of the metastatic cells (e.g., If developing brain metastases from a lung primary, inject into the lung). However, ectopic injections, where cells are inoculated directly into the circulation, are commonly used to avoid difficulties encountered during the initial stages of metastatic progression [6]. Inoculation through the tail vein allows cells to disseminate primarily to the lung and subsequent metastasis to the CNS, whereas inoculation injection into the intracarotid will see cells disseminate primarily to the brain and then the abdominal cavity [6]. Intracardiac injections allow systemic distribution of the cells throughout the body.

The material injected also plays a significant role in model development. Particular cell lines are commonly used due to the ease of culturing and manipulation; however, established lines have undergone clonal selection from years of culture and result in different genotypic and phenotypic profiles. Primary patient samples provide a more accurate representation of tumor heterogeneity. However, studies also face difficulties with obtaining enough samples or cells for experiments as well as long latency periods for tumor development.

Unfortunately, there still remain several limitations encountered with these models that have yet to be addressed, such as long latency periods from injection to metastatic development, a lack of robustness or reproducibility between samples and limited metastatic dissemination to certain organs. Also, there remains a dearth of models that can accurately replicate the clinical progression of the disease in its entirety [6]. Nonetheless, the use of in vivo models has contributed significantly to understanding the mechanisms involved in brain metastasis progression. These preclinical models also provide a vital tool for identification and screening novel therapeutic avenues. Here, we have developed a novel clinically relevant in vivo metastasis model utilizing primary patient samples of brain metastases from a primary lung cancer.

2 Materials

2.1 Tumor Processing and Primary Cell Culture

1. Dulbecco's phosphate-buffered Saline (PBS).
2. Liberase Blendzyme (0.2 Wunisch units/mL, Roche).
3. 1–2 in. sharp-ended scissors and 6–8 in. sharp-ended forceps
4. 70 μm nylon mesh cell strainer
5. RBC Lysis Buffer, ammonium chloride solution (STEMCELL Technologies).
6. 100 mm Ultralow attachment cell culture plates.
7. Neurocult media supplemented with 20 ng/mL EGF, 10 ng/mL FGF and 2 ug/mL Heparin (STEMCELL Technologies).
8. Incubator rocker.

2.2 Preparation for Injection

1. Dulbecco's phosphate-buffered saline (PBS).
2. Liberase Blendzyme (0.2 Wunisch units/mL, Roche).
3. Trypan blue.

2.3 Intracranial and Intracardiac Injections

1. 27 ga needle and syringe.
2. 1–2 in. sharp-ended scissors or razor for hair removal.
3. 70% ethanol wipes.
4. Low growth factor Matrigel (BD Biosciences).
5. Isofluorane anesthetic.
6. Rodent gas anesthesia instrument equipped with vaporizer, oxygen tank, charcoal scavenger filter containers for excess anesthetic, induction chambers, nose cones and hoses for anesthesia maintenance.
7. Heat pad for recovery.

2.4 Specimen Collection

1. 2.5% 2,2,2-tribromoethanol (Avertin): 5 g of Avertin powder is dissolved in 5 mL of tert-amyl alcohol and mixed/heated in a 50 °C water bath until dissolved. 97.5 mL of pre-warmed PBS is added for each 2.5 mL of Avertin-tert-amyl alcohol solution and subsequently filtered through a 0.22 μM filter. 5 mL aliquots are made into glass vacutainer tubes and stored away from light at 4 °C.

2. 1–2 in. sharp-ended scissors and 6–8 in. sharp-ended forceps.

3. 1 cc syringe and ½ cc syringe, per mouse.

4. Coronal mouse brain slicer matrix.

5. Razor blades.

6. 21 ga needles to attach to saline and formalin lines.

7. 10% formalin solution.

8. Ringer's lactate solution, 0.9% sodium chloride 1000 mL bag.

9. 1000 USP units/mL heparin sodium.

3 Methods

3.1 Cell Culture

Brain metastases (BM) originating from non-small cell lung carcinoma (NSCLC) primary samples are obtained from consenting patients, as approved by the Research Ethics Board at Hamilton Health Sciences.

1. Wash tumor in adequate PBS to remove residual blood.

2. Physically dissociate tumor with sterile scissors and forceps until a "slurry" solution of dissociated tissue and cells is achieved (*see* **Note 1**).

3. Transfer the slurry to a 50 mL falcon tube containing 15 mL PBS and with 200 μL Liberase Blendzyme. Incubate for 15 min at 37 °C on an incubator rocker (*see* **Note 2**).

4. The tissue lysate is filtered through the 70 μm cell strainer to remove undigested tissue is removed by a cell strainer.

5. The filtered lysate is centrifuged at $300 \times g$ for 5 min.

6. The supernatant is discarded, and the pellet is resuspended in 1 mL PBS. 4–12 mL RBC lysis buffer is added to lyse the red blood cells, incubating at room temperature for 5 min (*see* **Note 3**).

7. The cell suspension is diluted with 10 mL PBS and centrifuged at $300 \times g$ for 5 min.

8. The supernatant is discarded, the cell pellet resuspended in Neurocult media and transferred into a low binding cell culture plate (*see* **Note 4**).

9. BMICs are grown as tumorspheres that are maintained at 37 °C with a humidified atmosphere of 5% CO_2.

10. 1–2 weeks prior to in vivo work, BMICs are transduced with lentivector expressing GFP to allow easy separation of BMICs from mouse cells upon sample collection.

11. BMICs are maintained in Neurocult with the appropriate selective antibiotic for GFP positive cells for a minimum of 2 days prior to injection.

3.2 Preparation for Injection

1. BMIC tumorspheres are collected into a 15 mL Falcon tube and centrifuged at $300 \times g$ for 5 min.

2. The supernatant is discarded, and the cell pellet is resuspended in 0.5–1 mL PBS. The cell suspension is incubated with 10 μL Liberase in a 37 °C water bath for 5–10 min, until cell dissociation is achieved (*see* **Note 5**).

3. The cell suspension is diluted with 5 mL PBS, and centrifuged at $300 \times g$ for 5 min.

4. The supernatant is discarded, and the cell pellet is resuspended in 1 mL PBS.

5. Cell number and viability is assessed by Trypan blue.

6. Single cells are resuspended in PBS in a volume appropriate for each specific injection route.

    (a) For intrathoracic injections, 30 μL of PBS per mouse.

    (b) For intracardiac injections, 50 μL of PBS per mouse.

3.3 Surgical Techniques

Mouse preparation

1. All experimental procedures involving animals were reviewed and approved by McMaster University Animal Research Ethics Board (AREB).

2. All injection procedures are performed aseptically in a BSL2 hood, with all equipment sterilized prior to use.

3. Set up the rodent anesthesia machine as required.

4. Weigh the mouse, and place into the induction chamber (*see* **Note 6**).

5. Once the mouse is unresponsive (as determined by toe pinch), it is placed onto a platform for injection.

Intrathoracic (IT) injection

1. The mouse is first positioned to lie flat on the back, with the right forelimb taped away from the body to expose the chest.

2. The upper thoracic region is shaved and the area cleansed with 70% ethanol.

3. 30 μL cell suspension (5×10^5 cells) mixed with equal volume of undiluted growth factor reduced matrigel (*see* **Note 7**).

4. The cell suspension is injected with a 27 ga needle into the right lateral thorax, approximately 1 cm below the scapula. The needle is quickly advanced approximately 6 mm into the thorax and quickly removed upon release of the cell suspension (*see* **Note 8**).

5. After injection, the mouse is turned to lie on its left side to recover. Any signs of labored breath post-injection signal possible puncture of the pleural cavity and the mouse must be sacrificed immediately.

6. The mice are identified using ear notches following minimally invasive surgery, and monitored for recovery. Mice are monitored weekly for signs of illness, and then sacrificed upon signs of endpoint.

Intracardiac (ICa) injection

1. The mouse is positioned to lie flat on the back, with both forelimbs taped to the side to expose the chest.

2. The upper thoracic region is shaved and the area cleansed with 70% ethanol.

3. 100 μL cell suspension (2.5×10^5 cells) is injected with a 28 ga needle into the left ventricle over a 15 s time span, and the needle was slowly and carefully removed (*see* **Note 9**).

4. The mice are identified using ear notches following minimally invasive surgery, and monitored for recovery. Mice are monitored weekly for signs of illness, and then sacrificed upon signs of endpoint.

3.4 Specimen Collection

Immunohistochemistry

1. Anesthetize the mice with intraperitoneal injection of 25 mg/mL Avertin (18 μL/g of mouse weight).

2. Once the mouse unresponsive (as determined by toe pinch), it is laid on its back and forelimbs taped to the surface.

3. 20 μL of heparin is administered through intracardiac injection (left ventricle), to prevent blood clotting.

4. Saline is administered through intracardiac injection (left ventricle) for 10 min, to flush the body and remove blood.

5. Approximately 10 mL of 10% formalin solution is then administered slowly through intracardiac injection (left ventricle), to perfuse the body.

6. The brains and lungs (for IT and ICa injections) are harvested using scissors and forceps carefully so as to prevent damage to the organs, and placed into ~5 mL of 10% formalin for a minimum of 48 h at 4 °C.

7. The organs are then sectioned into 2 mm slices using a coronal mouse brain slicer matrix and razor blades.

8. Brain slices are placed into cassettes, and dehydrated in first 50% ethanol (5 min) and then 70% ethanol (minimum 24 h) in preparation for paraffin embedding and immunohistochemistry.

Recovery of BMICs

1. The mice are anesthetized with Avertin as above.

2. Once the mice are unresponsive (as determined by toe pinch), they are sacrificed by cervical dislocation.

3. Using scissors and forceps, the brains and lungs are aseptically removed and stored separately in 5–10 mL of PBS (*see* **Note 10**).

4. The organs are prepared and cultured similarly to patient brain tumor samples.

5. Cell pellets are cultured in Neurocult media supplemented with 20% FBS (*see* **Note 11**).

6. Over the subsequent 7 days, ~2 mL of cell suspension containing debris and dead mouse cells is discarded, and FBS concentration is gradually reduced to 5% FBS, then cultured in serum-free Neurocult.

7. Cells are maintained in Neurocult for a minimum of 2 days prior to any further experiments performed.

---

4 Notes

1. Take care to not expose the tumor "slurry" to air for a prolonged period of time to prevent unnecessary cell death.

2. Seal the falcon tube with parafilm to prevent leakage as the tube is rocking.

3. The amount of RBC lysis buffer added is dependent on the amount of RBCs present in the pellet, identified as a red layer above the cell pellet.

4. Neurocult media allows for continual propagation of cells with stem cell properties termed, brain metastasis initiating cells (BMICs). The amount of Neurocult media used is dependent on the size of the cell pellet. From our experience, a low or very high cell density is not conducive to the initial propagation of BMICs.

5. The addition of Liberase needs to be carefully monitored to prevent drastic reduction of cell viability. In our experience, each BMIC sample behaves different, and as such will require

differing amounts of Liberase and incubation time to achieve appropriate cell dissociation.

6. Mice are anesthetized using Isoflurane, induced at 5% and maintained at 2.5% for the surgical procedure.

7. Matrigel rapidly polymerizes at room temperature or above to form a gel, acting as an anchor and prevents tumor cells from diffusing throughout the thoracic cavity. It must be added immediately to the cell suspension and mixed gently before injection.

8. Any signs of labored breath post-injection indicate a possible blockage in or rupture of the lung and the mouse must be culled immediately.

9. Any signs of labored breath or no response post-injection indicates a possible blockage in or rupture of the heart and the mouse must be culled immediately.

10. The scissors are sterilized after each organ is removed and stored separately to prevent mixing of different BMIC populations.

11. Culturing the organs specimens in 20% FBS promotes the adherence and growth of the BMICs to the plate surface. Discarding of portions of the supernatant over the week helps to prevent the adherent cells from dying from external factors released as the residual mouse cells die.

References

1. Patchell RA (2003) The management of brain metastases. Cancer Treat Rev 29:533–540

2. Arvelo F, Sojo F, Cotte C (2016) Tumor progression and metastasis. Ecancermedicalscience 10:617. https://doi.org/10.3332/ecancer.2016.617

3. Singh M, Manoranjan B, Mahendram S, McFarlane N, Venugopal C, Singh SK (2014) Brain metastasis-initiating cells: survival of the fittest. Int J Mol Sci 15:9117–9133. https://doi.org/10.3390/ijms15059117

4. Liao WT, Ye YP, Deng YJ, Bian XW, Ding YQ (2014) Metastatic cancer stem cells: from the concept to therapeutics. Am J Stem Cells 3:46–62

5. Bos PD, Nguyen DX, Massague J (2010) Modeling metastasis in the mouse. Curr Opin Pharmacol 10:571–577. https://doi.org/10.1016/j.coph.2010.06.003

6. Daphu I, Sundstrom T, Horn S, Huszthy PC, Niclou SP, Sakariassen PO, Immervoll H, Miletic H, Bjerkvig R, Thorsen F (2013) In vivo animal models for studying brain metastasis: value and limitations. Clin Exp Metastasis 30:695–710. https://doi.org/10.1007/s10585-013-9566-9

Chapter 20

Cellular Magnetic Resonance Imaging for Tracking Metastatic Cancer Cells in the Brain

Katie M. Parkins, Ashley V. Makela, Amanda M. Hamilton, and Paula J. Foster

Abstract

Cellular magnetic resonance imaging (MRI) enables visualization of cells in vivo. This is accomplished by labeling cells with superparamagnetic iron oxide nanoparticles. Here, we describe the steps for labeling human cancer cells with iron for tracking them after injection into nude mice. We also provide details for validation of cell labeling, ultrasound guided intra-cardiac injection, and MRI.

Key words MRI, Cells, Iron oxide, Cell labeling, Cancer, Metastasis, Dormancy, Imaging, Methods

1 Introduction

Typically, magnetic resonance imaging (MRI) is used to image disease, and more specifically the anatomical manifestations of disease at a relatively late stage; for example, brain lesions caused by multiple sclerosis or cancerous tumours. Cellular MRI is an area under development that aims to image early stages of disease or early cellular events, for instance, the presence and location of infiltrating immune cells or metastatic cancer cells.

To visualize cells with MRI they must be labeled with an MRI detectable agent. The most commonly used agents are superparamagnetic iron oxide (SPIO) nanoparticles [1]. These are usually composed of three main components: an iron core, a polymer coating, and sometimes functional moieties. There are two general approaches to using iron particles for cellular MRI; iron particles can be administered intravenously (iv) with the goal of labeling phagocytic cells in situ (for inflammation imaging) [2] or they can be used to label cells in culture prior to the injection or transplantation of cells (cell tracking) [3]. The methods in this chapter will focus on labeling cells in culture for MRI cell tracking.

Sheila K. Singh and Chitra Venugopal (eds.), *Brain Tumor Stem Cells: Methods and Protocols*, Methods in Molecular Biology, vol. 1869, https://doi.org/10.1007/978-1-4939-8805-1_20, © Springer Science+Business Media, LLC, part of Springer Nature 2019

Iron oxide nanoparticles are available in a range of sizes and with different coatings. Standard SPIO are between 50 and 150 nanometers (nm) [1]. Ultra-small superparamagnetic (USPIO) iron oxide nanoparticles range in size from 15 to 35 nm and were developed to have a longer blood half-life after intravenous (IV) administration, allowing them to avoid rapid uptake by the phagocytic cells of the reticuloendothelial system thereby reaching deeper tissues [4]. SPIO and USPIO are typically coated with dextran or carboxy-dextran making them biocompatible and biodegradable. A version of SPIO (Feridex) was used in patients for years in North America but was taken of market in 2008 due to financial reasons. At the present time, there is one USPIO, a ferumoxytol (Feraheme), that is used off label for human imaging [5]. Micron-sized superparamagnetic iron particles are approximately 1–10 μm in diameter. These particles are coated with or embedded in a polystyrene or silicon bead and are inert and used only for preclinical research [6–9]. MPIO are often used for cell tracking because they are not degraded and because they have the highest iron content and therefore can be used to detect fewer cells, or cells containing fewer particles. This is important for tracking proliferative cells since the iron will be diluted as cells divide. For all types of particles, the coating can be functionalized with fluorescent dyes for in vivo or ex vivo fluorescent imaging validation.

Most cells can be labeled with these agents by simple coculture [10]. Iron is taken up by endocytosis and is compartmentalized in vacuoles in the cell cytoplasm. The amount of iron taken up per cell influences the detection by MRI. Ideally for MRI, cells should be labeled with as much iron as possible without affecting cell phenotype or function. A variety of cell types including stem cells [11], T-lymphocytes [12], dendritic cells [13–15], pancreatic islets [16], and cancer cells [17–19] have been labeled and tracked with MRI. Studies show minimal impact on cell function or phenotype at a wide range of iron loading levels. Depending on the cell type of interest it is important to check cell viability and the important and relevant cell functions for labeled versus unlabeled cells. The labeling efficiency is commonly evaluated using the Perls Prussian Blue stain for iron (Fig. 1) (described in Subheading 3).

To detect cells with MRI a pulse sequence that is sensitive to the effects of iron on the magnetic field is used. Iron-labeled cells appear as regions of signal void in these images. The size of the signal void in an image is much larger than the size of a cell because of an effect called "blooming." Images can be T2-weighted or T2*-weighted. T2*-weighted images are more sensitive to the presence of iron, therefore the blooming effect is larger and fewer iron-labeled cells can be detected [20].

Our lab has focused on imaging metastatic cancer cells using this technology. Cancer cells readily take up SPIO and MPIO. We have demonstrated that individual iron-labeled metastatic breast

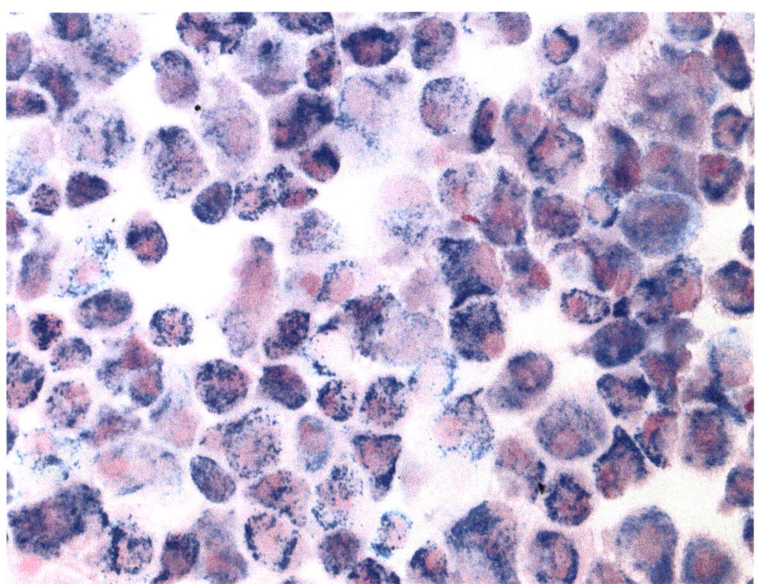

Fig. 1 Example of PPB staining of iron labeled cells

cancer cells can be detected by MRI in the mouse brain [21–24]. We have also shown that iron-labeled cancer cells can be tracked in vivo from their initial arrest in the brain to the development of brain metastases in mice [22]. In this paper, we used MRI to follow the fate of 231BR cells labeled with MPIO particles for 1 month; images were obtained on days 0, 4, 7 and then weekly. MPIO-labeled solitary cells, and small clusters of cells, were identified as regions of signal void in the brains of mice. These experiments revealed that at the end of the experiment (day 28 post cell injection) there were three distinct cell fates (Fig. 2). There was a population of cells that we called *"transient"* (94%); these were discrete signal voids in images on day 0 that disappeared over time (usually between days 0 and 7). There was a population of cells that we called *"non-proliferative"* (4.5%); these were discrete signal voids in images on day 0 that appeared unchanged on day 28. Finally, there was a population of cells that we called *"proliferative"* (1.5%); in images acquired on day 0 these cells appeared as discrete regions of signal loss but in images on day 28 they were replaced by a tumor.

We believe the *"nonproliferative"* cells represent dormant cancer cells. Evidence to support this comes from: (1) quantification of the signal voids—the degree of contrast of a signal void can be measured and can be related to iron content. Our measurements showed that the contrast of the voids, which represent *"nonproliferative"* cells, does not change over time. This suggests that these voids contain the same amount of iron at each time point. (2) iron-labeled 231BR cells rendered "apoptotic" have been shown to clear

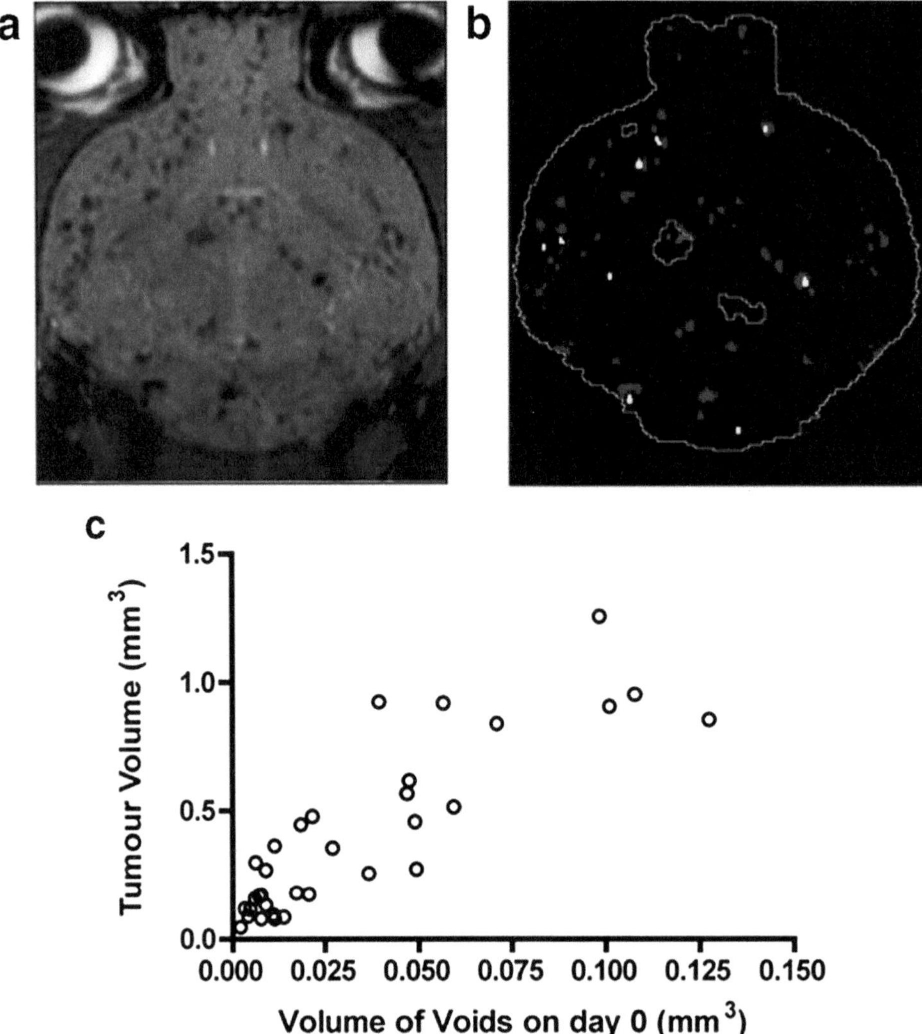

Fig. 2 Mouse ultrasound-guided intra-cardiac cell injections: Mouse (M) chest fur is removed and mouse is placed supine with limbs taped to the imaging platform (IP) and nose cone in place to maintain anesthetic. Hamilton syringe with 30G needle (N) is place into the holder (H). Holder can change needle trajectory (i.e., height, angle, x and y direction). Transducer (T) is placed in holder above mouse so when imaging platform is rotated, the mouse is turned toward needle and the transducer will be parallel to the long axis of mouse

much faster than control iron-labeled 231BR cells. This shows that dead/dying cells (our *"transient"* cell fate) are cleared efficiently from the brain. (3) We could isolate iron-positive cells from the mouse brain at endpoint, using a magnetic column, and found that these cells were viable. Our data provided the first in vivo evidence for dormancy in brain metastasis of breast cancer. The potential importance of tracking the fate of micrometastases and dormant cells in the brain cannot be overstated since it is the outgrowth of these that contributes to relapse.

2 Materials

2.1 Cell Labeling

1. MPIO beads (0.9 μm, conjugated with Flash Red, Bangs Laboratories, Inc., IN, USA). Store at 4 °C.

2. Dulbecco's Modified Eagle Medium (DMEM), 4.5 g/L D-glucose, L-glutamine. Store at 4 °C.

3. Hank's Balanced Salt Solution (HBSS), store at 4 °C.

4. Phosphate-buffered saline (PBS), store at 4 °C.

5. 0.25% trypsin-EDTA ($1\times$), store at 4 °C short term; or -20 °C long term.

2.2 Cell Viability Test

1. 0.4% Trypan blue solution (ThermoFisher scientific).

2. Hank's Balanced Salt Solution (HBSS), store at 4 °C.

3. 0.25% trypsin-EDTA (1X). Store at 4 °C short term; or -20 °C long term.

4. Hemocytometer.

2.3 Ultrasound-Guided Intra-Cardiac Injection

1. Vevo 2100-visualsonics ultrasound system.

2. 100 μL Hamilton syringe.

3. 30 G Hamilton needle.

4. 21 G Hamilton needle.

5. Vevo MicroMarker (QC3311). Formulation: *Gas:* Mixture of nitrogen and perfluorobutane. *Excipient* (non-active ingredient): polyethylene Glycol, Phospholipids and fatty acid. *Solvent;* Sodium Chloride 0.9% w/v in water.

6. Saline solution.

7. Ultrasound gel.

8. Nair for sensitive skin (if not using nude mice).

2.4 Perls Prussian Blue (PPB) Staining

1. Hemocytometer.

2. Methanol acetic acid solution (3:1).

3. 2% potassium ferrocyanide .

4. 2% hydrochloric acid.

5. Aluminum sulfate.

6. Nuclear fast red dye (Sigma, Cat# N8002-5G).

7. Whatman paper.

8. 70, 95, and 100% ethanol.

9. Xylene.

10. Cytoseal.

3   Methods

3.1   Cell Labeling

1. Start with a healthy <90% confluent flask of your cancer cell type of choice (*see* **Notes 1** and **2**).

2. Seed an adequate quantity of cells such that your tissue culture vessel of choice will be 50–70% confluent the following day. For example, for most cancer cell lines 2×10^6 cells would be an appropriate number of cells to seed in a T75cm^2 flask (*see* **Note 3**).

3. Supplement the cells with complete media (basal media containing fetal bovine serum (FBS)) and incubate overnight at 37 °C, 5% CO_2.

4. Prepare your labeling media by adding 50 μg Fe/mL of your nanoparticle of choice to a fresh aliquot of pre-warmed complete media (*see* **Note 4**).

5. Remove the old media and replace with the nanoparticle-containing complete media. Incubate for 24 h at 37 °C, 5% CO_2.

6. Remove nanoparticle-containing complete media and wash the cells with a large volume (e.g., 10 mL) of buffer (e.g., Hank's balanced salt solution, HBSS; or phosphate-buffered saline, PBS).

7. Remove the buffer and repeat for a total of three washes.

8. Remove buffer and add 1–2 mL of 0.25% trypsin. Incubate 3–5 min at 37 °C until the cells have dissociated.

9. Add 10 mL of complete medium to the disassociated cells to inactive the trypsin.

10. Collect the cells in a 15 mL falcon tube and centrifuge at 112 G for 5 min.

11. The resulting cell pellet of successfully labeled cells will appear brown. Remove the supernatant from the cell pellet and resuspend the cells in 12 mL of buffer (HBSS or PBS).

12. Repeat **steps 10** and **11** for a total of three washes.

13. After the final wash, resuspend the labeled cells in a lower volume of buffer for cell quantification and use in the desired application (i.e., experimental animal injection).

14. A subset of cells can be stained with Trypan blue and counted using a hemocytometer to assess concentration and cell viability (*see* **Note 5**).

15. To assess labeling efficiency, $2–3 \times 10^5$ cells can be adhered to a glass slide by Cytospin™ cytocentrifugation, according to the manufacturer's instructions. The slide can then be stained using the Perls Prussian blue staining protocol (*see* Subheading 3.3 below).

3.2 Cell Viability Test

1. Trypsinize adherent labeled cells, wash three times with HBSS.

2. Resuspend cells in HBSS. Pipette back and forth to evenly suspend the cells.

3. Pipette 10 μL of cells and place in an Eppendorf tube or microplate. Add 10 μL Trypan blue and mix with cell suspension.

4. Apply coverslip to hemocytometer first and then take 10 μL of the Trypan blue treated cell suspension and carefully introduce it in the space between the slide and the hemocytometer.

5. Place under microscope. Focus on the grid lines of the hemocytometer with a $10\times$ objective.

6. Use a hand tally counter to count the number of live, unstained cells in one set of 16 squares (1 of 4 corners of hemocytometer). Additionally, count dead, stained cells (which appear as dark; Trypan blue penetrates the membrane of dead cells). Move the hemocytometer to the next set of 16 corner squares and count until all four sets are counted.

7. To determine number of cells in original cell suspension:

$$\text{Total cells}/_{\text{mL}} = \text{Total cells counted} \times \frac{\text{dilution factor}}{\text{\# of squares}} \times 10{,}000 \ ^{\text{cells}}/_{\text{mL}}$$

*Multiply by 2 (dilution factor) to correct for the 1:1 dilution from the Trypan Blue addition

8. To determine viability, add together the live and dead cell count to obtain a total cell count. Divide the live cell count by the total cell count to calculate the percentage viability.

3.3 Perls Prussian Blue (PPB) Staining

1. Adherent cells can be either grown on chamber slides or adherent or non-adherent cells can be affixed to a glass slide using a Cytospin™ cytocentrifuge, according to the manufacturer's instructions. Cell staining for up to 10 slides can be performed in a coplin jar which has a 50 mL solution capacity.

2. Fix cells in a 3:1 (v/v) solution of Methanol/Acetic acid solution for 5 min at room temperature (*see* **Note 6**).

3. Remove the fixative and rinse with water. Set aside.

4. Prepare the PPB solution fresh and use within 1 h (*see* **Note 7**). A 2% (w/v) potassium ferrocyanide (potassium ferric ferrocyanide) solution must be dissolved fresh in laboratory grade water. 2% hydrochloric acid can be made in advance. Combine equal volumes of ferrocyanide and hydrochloric acid and filter through Whatman paper before use.

5. Stain the slide in the freshly prepared PPB solution for 15 min at room temperature (*see* **Note 8**).

6. Remove the stain and dispose of in a waste bottle.

7. Rinse the slide with dH_2O.

8. Counterstain in nuclear fast red for 2–5 min at room temperature. This counterstain can be prepared in advance by dissolving 5 *g* aluminum sulfate in laboratory grade water, then add 0.1 g nuclear fast red dye and slowly heat to boil and cool. Filter out any undissolved stain using Whatman paper and add a grain of thymol as a preservative. Nuclear Fast Red can be reused and stored for up to 1 year, but should be filtered using Whatman paper prior to each use.

9. Rinse with dH_2O.

10. Dehydrate through 2 min exchanges of increasing concentrations of ethanol (i.e., 70%, 95%, 100%).

11. Clear in xylene (at least 2 exchanges and leave in final step for at least 5 min).

12. Immediately coverslip with Cytoseal (or any xylene-based mounting medium) and dry in a fume hood.

13. Once dry the slide can be imaged by brightfield microscopy. Nanoparticles will appear deep blue. Nuclei will appear red and cell bodies will appear pale pink.

3.4 Ultrasound-Guided Intra-Cardiac Injection

1. Prepare cells as in Subheading 3.1 to yield required number of cells per mouse in 85 µL HBSS, multiplied by number of injections (*see* **Note 9**). For mouse studies of breast cancer brain metastasis using human cancer cells 6–7 week old nude (nu/nu) mice are used.

2. Prepare ultrasound microbubbles (i.e., Vevo MicroMarker Contrast Agent). Using a 21G needle attached to syringe, draw up 700 µL saline. Inject saline directly into glass vial.

3. Remove syringe and leave needle in rubber top to vent for 5 min. Remove needle and shake container gently for 10 s. Add 15 µL of microbubbles per 85 µL of cells in HBSS to yield 100 µL cell mixture per mouse/injection (*see* **Note 10**).

4. If not using nude mice, remove fur from the mouse thorax (i.e., hair removal cream) (*see* **Note 11**). Anaesthetize mouse in 2% isoflurane in O_2 in an induction chamber.

5. Maintain mouse under anesthetic via a nose cone for entire procedure; place mouse on its back (supine) on an imaging platform. Tape limbs down so mouse torso is taut and exposed (Fig. 3).

6. Log onto ultrasound system and utilize cardiology mode application preset. Initialize transducer with center transmit of 40 MHz and mount to holder above mouse so transducer is parallel to long axis of mouse.

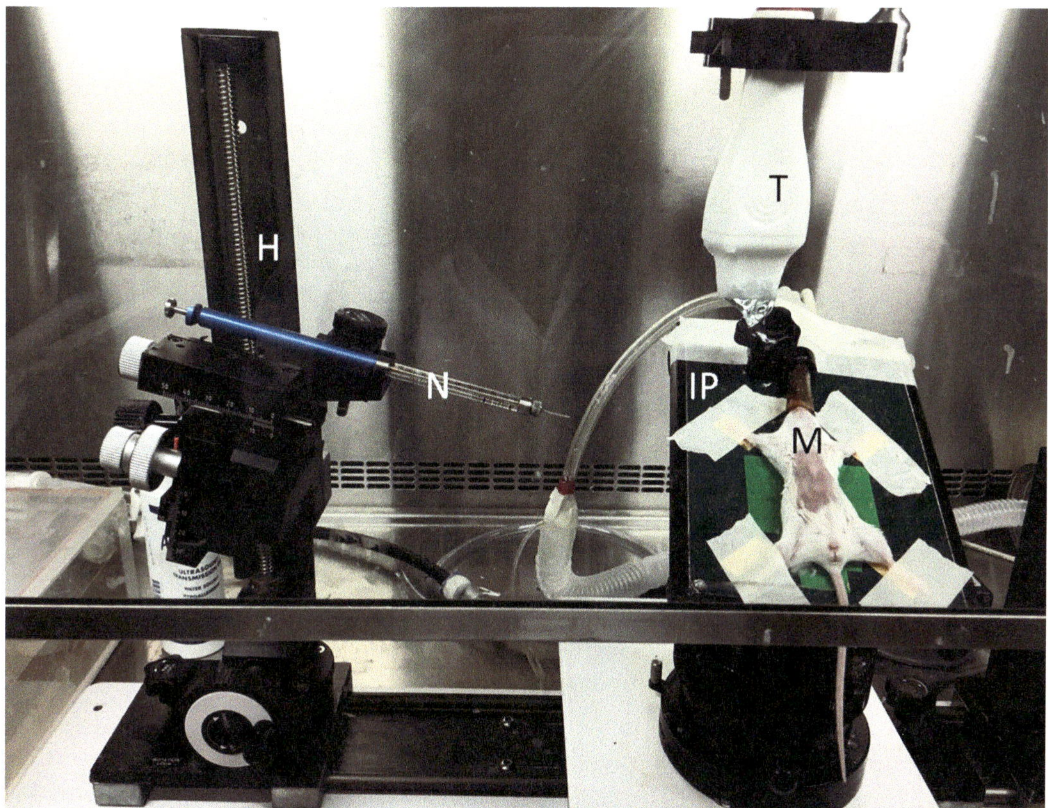

Fig. 3 MR tracking of tumor cell fate. (**a**) Axial MR image of a mouse injected with 100,000 MDA-MB-231BR/EGFP cells on day 0. (**b**) Cell-fate map for cells shown in a. The brain is outlined in white. Blue areas correspond to "transient tumor cells" that are visible on day 0 but disappear by day 28. Green outlines correspond to boundaries of tumors visible on day 28. Red areas correspond to cells that reside within the area circumscribed by the tumor boundary at day 0, and therefore the population of cells that give rise to the tumors ("proliferating tumor cells"). One tumor outlined in green is associated with cells (red) within the slice shown. The other two tumors are associated with cells that appear on a different MR slice (data not shown). Yellow areas correspond to nonproliferating cells that are visible on days 0 and 28. (**c**) Correlation of initial cell number with end tumor volume. The tumor volume at day 28 is plotted against the volume of signal voids circumscribed by the tumor on day 0. This void volume is proportional to the number of cells associated with a given tumor. Larger tumors arise from areas with more cells on day 0, indicating an effect of local cell dose on tumor size. From ref. 22

7. Using a 100 μL Hamilton syringe equipped with a 30 G needle tip, draw up 100 μL of the cell/microbubble mixture into syringe. Secure syringe into holder.

8. Turn mouse to face syringe, tipping mouse down (~45°) and slightly to the right (*see* **Note 12**). Apply a thick layer of ultrasound gel to the chest of the mouse.

9. Lower ultrasound transducer until there is contact with gel (do not make contact with mouse chest). Move imaging platform so transducer is over the left side of the mouse. Begin

moving mouse back to center and when heart is visualized stop here; this is the left ventricle.

10. Adjust needle in x and y direction while between mouse and transducer, so entire needle length and tip are appropriately visualized (seen as hyperechoic line by ultrasound).

11. Adjust needle (angle, height) so it approaches just below the mouse xiphoid bone at ~20–30°. Advance the needle until it penetrates the skin and muscle (will visualize "bounce back of chest" on ultrasound).

12. Retract needle slightly (~1 cm), adjust appropriately (decrease angle, decrease height). Advance needle, making adjustments where necessary to enter left ventricle at an almost parallel approach.

13. Once needle tip is within the left ventricle, slowly inject cells. This will be confirmed by an enhancement (hyperechoic by ultrasound) of blood in left ventricle due to microbubbles (*see* **Note 13**).

14. Retract needle, raise transducer, and clean/remove mouse for recovery.

3.5 Magnetic Resonance Imaging

1. MRI can be performed at any magnetic field strength; the higher the field strength the more sensitive to iron-labeled cells. Our lab has successfully imaged cells at 1.5 Tesla (T), 3.0 T, and 9.4 T.

2. A few different imaging pulse sequences can be used: T2-weighted spin echo, T2*-weighted gradient echo or a balanced steady state free precession (bSSFP) imaging sequence (this is what we use).

3. Anesthetize mice using 2–2.5% isoflurane in oxygen (2 L/min flow rate) in a chamber before transferring to the radiofrequency (RF) coil and nose cone. Use 1–1.5% isoflurane for maintenance of anesthesia during imaging. Attach scavenger for exhaled gas. We use an activated carbon charcoal filter system for passive scavenging.

4. Position the mouse head inside of a RF coil. The smaller the coil the better the sensitivity. In our lab, we use a custom-built solenoidal RF coil which has an inner diameter of 1.5 cm. A volume or surface coil can also be used.

5. Body temperature should be maintained during imaging. This can be done simply by placing warmed bags of water or saline near the mouse and outside of the RF coil. Alternatively, a warm air feedback system can be used with active feedback.

6. Set imaging parameters. For bSSFP (called FIESTA on GE MRI systems and TrueFISP on Siemens systems) the most important imaging parameters for maximizing the contrast

due to iron-labeled cells are the echo time, repetition time, receiver bandwidth, and spatial resolution. See Ramadan et al. [25]. Total scan time should be under 1 h for live mice.

4 Notes

Notes on Cell Labeling

1. Freshly thawed cells should be grown for at least a week before labeling to optimize cell labeling efficiency and cell viability.

2. Cells should ideally be grown without the use of antibiotics if the desired application is experimental animal model injection. The presence of antibiotics can mask low-level bacterial contamination that should not be transmitted to experimental animals.

3. Trypsin can be inactivated by residual FBS left on the cells after media removal. It can also be adversely affected by the presence of Mg^{2+} or Ca^{2+} in the rinsing buffer (HBSS or PBS). If cell disassociation is not accomplished within 5 min at 37 °C, consider rinsing with a greater volume of buffer (without Mg^{2+} or Ca^{2+}) prior to trypsinization or using a greater volume of trypsin. If problems still persist, try switching to alternative disassociation agents such as TrypLE.

4. Most nanoparticles label well at a concentration of 50 μg Fe/mL, however, the product information for your nanoparticle of choice should be referenced when choosing the appropriate concentration.

5. Automated methods of cell quantification and viability assessment should not be used for nanoparticle labeled cells. These assessments have issue differentiating a well labeled cell from a Trypan blue-stained dead cell.

Notes on PPB Staining

6. In vitro labeled cells can be fixed with a 3:1 (v/v) solution of Methanol/Acetic acid. Other fixatives may be used such as neutral buffered formalin but dichromate fixatives should be avoided.

7. The PPB solution will begin to react and turn blue after 1 h. Solution that is more than 1 h old should never be used.

8. Slides can be left in PPB solution for up to 30 min; however, this long incubation should not be necessary for well-labeled cells. Staining should not be extended beyond 30 min.

Notes on Cell Injection

9. Prep for 2 more mice than necessary to ensure there is enough of the cell mixture.

10. Keep the cell/microbubble mixture on ice and slightly mix before each drawing of cells. Clumping of cells could cause emboli, stroke, or death of the animal after intracardiac injection.

11. If not using nude mice, make sure as much hair as possible is removed as this could impair image quality (needle tip/heart chamber visualization) by ultrasound.

12. When positioning mouse and transducer to achieve best visualization of left ventricle, left ventricle should be larger vs other chambers of heart. Tipping (rotating) mouse along its long axis may help with a better view.

13. If upon injection there is a single hyperechoic foci, the injection may be in the heart wall. Confirm placement of needle tip and try again.

References

1. Di Marco M, Sadun C, Port M, Guilbert I, Couvreur P, Dubernet C (2007) Physicochemical characterization of ultrasmall superparamagnetic iron oxide particles (USPIO) for biomedical application as MRI contrast agents. Int J Nanomedicine 2(4):609

2. Neuwelt A, Sidhu N, Hu CA, Mlady G, Eberhardt SC, Sillerud LO (2015) Iron-based superparamagnetic nanoparticle contrast agents for MRI of infection and inflammation. AJR Am J Roentgenol 204(3):W302–W313. https://doi.org/10.2214/AJR.14.12733

3. Li L, Jiang W, Luo K, Song H, Lan F, Wu Y, Gu Z (2013) Superparamagnetic iron oxide nanoparticles as MRI contrast agents for non-invasive stem cell labeling and tracking. Theranostics 3(8):595–615. https://doi.org/10.7150/thno.5366

4. Zhao X, Zhao H, Chen Z, Lan M (2014) Ultrasmall superparamagnetic iron oxide nanoparticles for magnetic resonance imaging contrast agent. J Nanosci Nanotechnol 14 (1):210–220

5. Vasanawala SS, Nguyen KL, Hope MD, Bridges MD, Hope TA, Reeder SB, Bashir MR (2016) Safety and technique of ferumoxytol administration for MRI. Magn Reson Med 75(5):2107–2111. https://doi.org/10.1002/mrm.26151

6. Hinds KA, Hill JM, Shapiro EM, Laukkanen MO, Silva AC, Combs CA, Varney TR, Balaban RS, Koretsky AP, Dunbar CE (2003) Highly efficient endosomal labeling of progenitor and stem cells with large magnetic particles allows magnetic resonance imaging of single cells. Blood 102:867–872

7. Shapiro EM, Sharer K, Skrtic S, Koretsky AP (2006) In vivo detection of single cells by MRI. Magn Reson Med 55(2):242–249

8. Rohani R, de Chickera SN, Willert C, Chen Y, Dekaban GA, Foster PJ (2011) In vivo cellular MRI of dendritic cell migration using micrometer-sized iron oxide (MPIO) particles. Mol Imaging Biol 13(4):679–694

9. Bernas LM, Foster PJ, Rutt BK (2010) Imaging iron-loaded mouse glioma tumors with bSSFP at 3 T. Magn Reson Med 64(1):23–31

10. Frank JA, Anderson SA, Kalsih H, Jordan EK, Lewis BK, Yocum GT, Arbab AS (2004) Methods for magnetically labeling stem and other cells for detection by in vivo magnetic resonance imaging. Cytotherapy 6:621–625

11. Sykova E, Jendelova P (2007) In vivo tracking of stem cells in brain and spinal cord injury. Prog Brain Res 161:367–383

12. Shapiro EM, Medford-Davis LN, Fahmy TM, Dunbar CE, Koretsky AP (2007) Antibody-mediated cell labeling of peripheral T cells with micron-sized iron oxide particles (MPIOs) allows single cell detection by MRI. Contrast Media Mol Imaging 2(3):147–153

13. Zhang X, de Chickera SN, Willert C, Economopoulos V, Noad J, Rohani R, Wang AY, Levings MK, Scheid E, Foley R, Foster PJ, Dekaban GA (2011) Cellular magnetic resonance imaging of monocyte-derived dendritic cell migration from healthy donors and cancer patients as assessed in a scid mouse model. Cytotherapy 13(10):1234–1248

14. de Chickera S, Willert C, Mallet C, Foley R, Foster P, Dekaban GA (2012) Cellular MRI as a suitable, sensitive non-invasive modality for correlating in vivo migratory efficiencies of different dendritic cell populations with subsequent immunological outcomes. Int Immunol 24(1):29–41

15. Dekaban GA, Snir J, Shrum B, de Chickera S, Willert C, Merrill M, Said EA, Sekaly RP, Foster PJ, O'connell PJ (2009) Semiquantitation of mouse dendritic cell migration in vivo using cellular MRI. J Immunother 32(3):240–251

16. Jirak D, Kriz J, Strzelecki M, Yang J, Hasilo C, White DJ, Foster PJ (2009) Monitoring the survival of islet transplants by MRI using a novel technique for their automated detection and quantification. MAGMA 22(4):257–265

17. Foster PJ, Dunn EA, Karl KE, Snir JA, Nycz CM, Harvey AJ, Pettis RJ (2008) Cellular magnetic resonance imaging: in vivo imaging of melanoma cells in lymph nodes of mice. Neoplasia 10(3):207–216

18. Perera M, Ribot EJ, Percy DB, McFadden C, Simedrea C, Palmieri D, Chambers AF, Foster PJ (2012) In vivo magnetic resonance imaging for investigating the development and distribution of experimental brain metastases due to breast cancer. Transl Oncol 5(3):217–225

19. Ribot EJ, Foster PJ (2012) In vivo MRI discrimination between live and lysed iron-labelled cells using balanced steady state free precession. Eur Radiol 22(9):2027–2034. https://doi.org/10.1007/s00330-012-2435-0

20. Makela AV, Murrell DH, Parkins KM, Kara J, Gaudet JM, Foster PJ (2016) Cellular imaging with MRI. Top Magn Reson Imaging 25 (5):177–186

21. Heyn C, Ronald JA, Mackenzie LT, MacDonald IC, Chambers AF, Rutt BK, Foster PJ (2006) In vivo magnetic resonance imaging of single cells in mouse brain with optical validation. Magn Reson Med 55:23–29

22. Heyn C, Ronald JA, Ramadan SS, Snir JA, Barry AM, MacKenzie LT, Mikulis DJ, Palmieri D, Bronder JL, Steeg PS et al (2006) In vivo MRI of cancer cell fate at the single-cell level in a mouse model of breast cancer metastasis to the brain. Magn Reson Med 56:1001–1010

23. Murrell DH, Zarghami N, Jensen MD, Dickson F, Chambers AF, Wong E, Foster PJ (2016) MRI surveillance of cancer cell fate in a brain metastasis model after early radiotherapy. Magn Reson Med. https://doi.org/10.1002/mrm.26541

24. Parkins KM, Hamilton AM, Makela AV, Chen Y, Foster PJ, Ronald JA (2016) A multimodality imaging model to track viable breast cancer cells from single arrest to metastasis in the mouse brain. Sci Rep 6:35889. https://doi.org/10.1038/srep35889

25. Ramadan SS, Heyn C, Mackenzie LT, Chambers AF, Rutt BK, Foster PJ (2018) Ex-vivo cellular MRI with b-SSFP: quantitative benefits of 3T over 1.5 T. MAGMA. 21 (4):251–259. https://doi.org/10.1007/s10334-008-0118-2

INDEX

A

Animal stereotaxic surgery 202, 204

B

Boyden chamber 94–98, 115
Brain tumour stem cells (BTSCs) .... v, 4–6,
    11–19, 80–84, 118

C

Cancer stem cells (CSCs) ....v, 3–6, 11,
    12, 19, 70, 79, 85, 117–125, 128, 155, 156, 189,
    190, 232
Cell differentiation 85–87
Cell labeling 239, 240,
    243, 244, 249
Cell viability test.....................................243, 245
ChIP-reChIP 148, 149
ChIP-sequencing (ChIPseq) 150, 151
Chromatin Immunoprecipitation (ChIP)
    analysis 149, 150
Copy number analysis 49
CRISPR-Cas9 technology
    gRNA library amplification 169–171,
        173, 174, 176, 177, 183
    large scale CRISPR gRNA lentivirus
        production 174, 175
    lentivirus multiplicity of infection (MOI)
        determination 176, 177, 183, 184
Cryogenic preservation.....................................15, 16, 18
CyTOF (Mass Cytometry) 155–167

D

3D chemogradient chamber....................................95, 99
Determination 176, 177
DNA methylation profiling 38, 39,
    44, 45, 48
3D tumor spheroid invasion 108, 112, 114

E

Electroporations...................................... 171, 173, 174,
    210, 212, 215, 217, 220, 222, 224, 227, 228
ES cell targeting 208, 210,
    213, 227, 228

F

Flow cytometry
    extracellular staining 72
    intracellular staining.................................72, 73
    setting up for analysis................................. 73
    setting up for sorting 73, 74
Fluoroblok™ tumor invasion for coculture 107
Fluoroblok™ tumor invasion for co-culture 107, 109

G

Gene expression profiling 37, 38, 43, 44
Glioblastoma (GBM)...................................... 11–20, 37,
    46, 52, 70, 82, 83, 85, 94, 118, 119, 123, 124,
    155–167, 197

H

Hanging-drop/3-dimensional collagen migration
    assay...................... 128, 129, 132, 133, 135, 139

I

Immunohistochemistry 236, 237
Implanting spheroids 132, 133
Intracardiac (ICa) injection 232, 233,
    235, 236, 250
Intracranial and intracerebellar injections................... 129,
    130, 134, 136, 138
Intrathoracic (IT) injection 235, 236

L

Laminin coating 26, 128, 131, 133, 134
Lentiviral transduction............................... 30, 128, 129,
    133, 134, 139, 170
Limiting dilution analysis 19, 80–83, 195

M

Magnetic resonance imaging (MRI)................... 239, 247
Medulloblastoma (MB) 5, 23–33,
    37–39, 46, 48, 52, 70, 128

N

Nanopattern 95, 100–102
NanoString nCounter.....................................57–66

Sheila K. Singh and Chitra Venugopal (eds.), *Brain Tumor Stem Cells: Methods and Protocols*, Methods in Molecular Biology,
vol. 1869, https://doi.org/10.1007/978-1-4939-8805-1, © Springer Science+Business Media, LLC, part of Springer Nature 2019

Neural stem cells (NSC) 2, 3, 11,
    17, 30, 89, 90, 118, 194, 199
Neurosphere formation assay 30, 31
Nucleic acid extraction
    DNA isolation from FFPE tissue 42, 43
    DNA isolation from frozen tissue 41, 42
    RNA extraction 40, 41

O

Organotypics 107, 111, 112, 115
Osmotic mini-pumps 197–199, 202–204

P

Pathway analysis 38, 40, 49, 50
Pathway analysis software 40
Perfusion and fixation of NOD SCID
    brains 128, 130, 135
Perls prussian blue (PPB) staining 240, 243–246
Pooled CRISPR knockout screen 170,
    171, 177–183
Pre-processing and clustering software 39, 40
Proliferation assay 176, 190–193

R

Recombineering 208–210, 212, 216

S

Self-renewal assays 79–84, 191, 193
Soft agar assay 80–82
Spot assay 106, 108, 109
2-Stage generation of neural precursor cells
    from hESCs 128, 131
Standard ChIP protocol 145–148

U

Ultrasound-guided intra-cardiac injection 242, 243,
    246–248

V

Vector cloning 209, 210, 229

W

Wound healing 96, 101, 103